"From this distance in time, it seems obvious: After more than a century of dramatic, seemingly preordained expansion, the Lakota were about to face inescapable catastrophe when their food source, the buffalo, disappeared. Not so obvious, especially today, is what a society about to confront such changes is supposed to do about it."

—Nathaniel Philbrick, *The Last Stand*

COME QUICK AND BRING THE VILLAGE

Technology, Food, and Freedom

By Patrick Finney

BENTEEN. COME ON. BIG VILLAGE. BE QUICK. BRING PACKS. Final orders from Lieutenant Colonel George Armstrong Custer to Captain Frederick Benteen shortly before Custer commenced his advance at The Battle of the Little Bighorn. Custer realized he had badly underestimated the size of the village and needed more men. Benteen delayed, ignoring the explicit urgency in the message. Exposed and desperate, Custer attacked without reinforcements. Within the hour, the entirety of his immediate command, some two hundred men, would be crushed by nearly two thousand Lakota, Arapaho, and Cheyenne warriors who refused to accept that their way of life was all but over.

TABLE OF CONTENTS

FOREWORD

IT'S OCTOBER 2019. I'm standing on Reno-Benteen Hill at The Little Bighorn Battlefield National Monument, looking down at the Little Bighorn River and the Bighorn Mountains in the distance. I had been planning this trip for a year and had been reading about the battle and its critical role in American history for many years before that. Anyone who finally accomplishes a significant goal can tell you that often the feelings you anticipate you will have never actually materialize. It's rarely how you imagined it would be. What you thought would be a salient, but predictable event in your life ends up being something different, something unexpected, but ultimately brings the entire journey into a sharper focus and a clearer understanding.

It had been an interesting trip before I ever arrived in Montana. Being an early riser, I had planned to leave one Saturday morning before dawn, hoping to make it to Sioux Falls by early afternoon. This would leave plenty of time to review my itinerary and make any last-minute adjustments before I headed west. I had nine days off from work, but four full days would be spent just getting there and back. It wasn't going to be a vacation in the traditional sense. My plans were fairly aggressive, and I was hoping to travel three thousand miles. I was writing a book after all, and everything in my life was geared to that end.

I had a dispute with Enterprise on Friday morning concerning a reservation I had made months in advance. The bottom line was that I decided I would drive my own vehicle instead and save quite a bit of money in the process. I arrived home after work that evening feeling very good about the seven hundred dollars I wouldn't have to spend on a rental car. But I couldn't sleep that night. I kept waking up, wondering if my old work truck with 250,000 miles on the odometer that I had beat the crap out of for ten years farming and landscaping, would carry me through. A breakdown in the middle of South Dakota or Wyoming or Montana would surely ruin the trip, given my tight schedule. Because of financial and time constraints, there wouldn't be a second chance. Therefore, I decided to swallow my pride, and I arrived at Enterprise before they opened on Saturday morning and, quite impatiently, waited for my turn in line. The man behind the counter told me he wasn't sure if he would have anything available. I felt a tinge of panic creep up from my stomach. I was already four hours behind schedule and might have to roll the dice and drive my own truck anyway? What was the point of waiting? I asked about the Durango I had reserved. It was gone. He said to come back in a few hours, and he would see what he could do. A few hours? The day was slipping away from me, and it was my own fault. I told him I would take any car he had available—anything with four wheels. A year of planning, and now I couldn't even get on the road.

I returned an hour later, hoping to get lucky. I was told that he had one car available. Fantastic, I thought. Just a little hiccup and things are back on track. He informed me that it was a 2019 Dodge Caravan. What? A minivan? I asked if he had any other car than that one, anything at all. He said there was absolutely nothing else available. Was he sure? Yes, he was sure. I had to bite my tongue before I nearly asked if I could rent his own car instead—whatever he drove—anything was better than a minivan. I looked at the clock on the wall behind him, and realized I had to accept my fate. I was running out of time, and the nearest rental franchise would likely close before I got there. I thanked him for his help and got underway as quickly as possible.

As I drove north towards Jefferson City, I took stock of my minivan. It was very big, too big for one person, and the luggage looked ridiculous behind me as it barely made a dent in the cargo space. After driving for an hour or so, I came to the sobering realization that as an unmarried, white male in his late thirties, driving a rented minivan across the country, I fit every profile on the books. Pedophile. Serial Killer. Terrorist. Any of them would have been perfectly applicable. The FBI, US Marshalls, and Secret Service followed me all the way to Sioux Falls. At one point, I turned on NPR and learned that Homeland Security was holding a press conference regarding a new and immediate threat in the Midwest. "Sounds about right," I thought as I turned the dial. A Goddamned minivan. Traveling from town to town, I got into the habit of repeating a mantra to every waitress, hotel clerk, and gas station attendant that would listen: "It's a rental. It's all they had." No one believed me.

After a few days, I decided that as long as the minivan didn't break down and made it through the trip, I could live with it. The gas mileage was better than my truck ever would have gotten, so there was at least one silver lining. Regardless, the journey had finally begun, and I had other things to worry about. One of the most important places I planned to visit was the site of the Wounded Knee Massacre on the Pine Ridge Reservation in South Dakota. I wanted to see the inevitable conclusion of what happens when a state entity comprehensively usurps a people's agency. I wanted to see what the Seventh Cavalry did to civilians to compensate for what they could not do on the battlefield. There was a palpable vibe at the massacre site, an eerie sense of emptiness as so much history is forever frozen in time in that valley, but only for those of us who care to look, who want to understand how such an egregious act could be committed by our government against a people and a culture in the way of progress, who simply refused to give up their way of life. I'm glad I went because it was important to experience that side of our country's legacy, if only in a brief and detached way.

   While in the Oglala College Bookstore, I purchased a book entitled, *After Custer: Loss and Transformation in Sioux Country* by Paul L. Hedren. A man waiting in line behind me saw the book and laughed quite loudly, "After Custer, huh? Well, he went to hell! I know 'cause I been there!" George Custer has received as much attention as any other figure in American history during the nineteenth century. Rightly or wrongly, many Americans, including myself, find his life and his final battle endlessly fascinating. Custer was an instrument of the nation, sent by the nation, to do the nation's bidding. If he is indeed in hell, it could be anyone of us. I have often wondered what Custer was thinking when he finally saw the village in its totality. With his prolific combat experience during the Civil War, and his less extensive but still reasonable track record of Indian fighting, he must have known quite clearly, and immediately, the severity of the situation. His expedient orders to Benteen for support underscores this, but what I am looking for, what I am hoping to explore in this book, is something deeper. Did Custer understand, with any complexity, the historical moment he was now in? Did he worry that maybe the chickens had come home to roost, the hornet's nest had finally been kicked, and perhaps he was about to become the fall guy for nearly four hundred years of oppression—from Christopher Columbus to John Chivington?

   The feelings I had while standing on that hill looking down at the Little Bighorn River were much different than I had anticipated. For years I thought that I would have given anything to have been there for the battle, to have had a front-row seat, so to speak—the very seat that Curly, Custer's youngest Crow scout had, as he watched untrammeled and with complete horror, while his comrades were killed and mutilated during the greatest Indian victory in the West. I wanted to see Gall and Crazy Horse in battle, to see exactly when Custer died and how he comported himself at the end. I wanted to see Benteen's face when he received Custer's final, desperate orders, and I wanted to see the great exodus of eight thousand men, women, and children leaving the battlefield and heading for the most perilous of futures. But as I looked at the river and the Bighorn Mountains in the distance, the Crow's Nest, Sharpshooter Ridge, and the timber where Reno retreated, I could not help but be struck by the sheer beauty of the battlefield and the surrounding landscape. Even with the highway, the houses, and modernity, one can still picture how it must have looked in June 1876. I felt viscerally that I no longer wanted to be in Curly's shoes. I wanted to see the river and the mountains and the hills and the wildlife on June 24th, 1876. I wanted to see the village the day *before* Custer attacked, to see eight thousand people, a thousand lodges, and twenty thousand ponies grazing in the valley, the last great gathering of Plains Indians spread out for miles living as they had for two centuries. I wanted to see what true freedom and agency looked like for those people, arguably the last free people that our nation ever knew, clinging to a way of life that could not last. The lessons from that gathering and the subsequent battle reverberate to this day and could not be more pertinent to our culture than any other historical event.

Perhaps a hundred years from now, a high school kid will stumble across this book rotting away in an abandoned outhouse somewhere, where it likely came in handy when somebody ran short on Charmin. I hope that kid reads a few pages, and I hope he understands then what I understand about the Little Bighorn now. I hope he can envision what life was like for us, the few recalcitrant citizens who stubbornly clung to freedom and agency in our lives before we surrendered to a fully-automated, tech-driven world and curated life experience. The Lakota fought heroically to preserve their way of life against forces they could never hope to defeat in the long run. What about us—21st-century citizens? What resistance are we offering? Are we fighting at all? Or are we simply asking for, even impatiently demanding, the very world that will ultimately destroy our humanity?

1

COLTER'S CALLING

The last of the darkness was beginning to fade as he squatted beside the river and cupped water into his mouth. It was so cold that it stung his teeth, but it tasted clean and pure. He scanned the bluffs. There was no movement, but the dawn twilight was fast approaching. He quickly dumped water onto the fire and smothered it. A few wafts of smoke escaped, and he watched them float above his head and into the sky. He cursed himself for making a fire, but it had been brutally cold, so cold he thought the air itself would freeze into a block of ice.

He rechecked his supplies: extra moccasins, powder horn, canteen, steel for forming balls, flintlock long rifle, and hunting knife. He watched the last of the fire die out and thought about the wisdom of the trip. He needed the traps. Even if one or two was rusted beyond use. They were his livelihood. Potts had thrown his over as well so perhaps he could score double. Potts would like that. He thought about waiting until dusk and traveling at night, but the cloud cover that had lasted all week prevented any glow from the moon at night. The Blackfoot traveled at night too. He thought they might keep to their winter quarters with the relentless cold, but if he stumbled into a war party in the darkness, his chances of survival were slim.

He shoved off and began paddling downstream. He fell into an easy rhythm and began making good headway. He was still twenty-five miles from the fork. A dull unease came over him. How ridiculous would it be for him to get this far only to be ambushed so close to the traps? He figured he could make it in six hours if he paddled long and hard. The sun was rising behind him, but its rays offered little protection against the cold. After three miles, he caught sight of a white bear about two hundred yards ahead of him in the middle of the river. It watched him approach with what he hoped was a wearisome indifference. The bear was lean, too lean, and it wasn't moving. He pulled ashore onto a small gravel bar about a hundred yards upstream from the bear. He rammed a ball down his rifle and checked the powder. He set the hammer and took aim. The bear swung his head low and grunted and did not move. He squeezed the trigger and braced for the recoil. The smoke from the blast briefly blinded him, but he caught a glimpse of the bear bounding out of the river and into the woods along the riverbank. He quickly reloaded and set the hammer. He paused for a minute or so with the rifle leveled along the riverbank. There was no movement. He quietly pushed off into the river and worked his way down to the bank where he had last seen the bear. He stopped thirty yards short of where the bear had left the river, leveled his rifle, and pulled ashore. He searched for hair, blood, anything that might indicate a hit. He hoped to find nothing. He fired more in an attempt to get the bear moving out of the area than in any real hope that he would harvest some meat.

A few years before, he watched a pair of trappers fire two balls into a charging bear with what appeared to be no effect. One was killed outright, and the other was gored beyond recognition. He jumped into the river and fired a ball while the first man was being attacked. He reloaded as fast as he could while the second man was being mauled and fired again. He believed both balls passed through the bear's brisket. He pulled the second man ashore and dressed his wounds, but there had been too much blood loss, and a few minutes later the man was dead.

It was very unlikely that he killed the bear with one ball from that range. A superficially wounded bear could be on him in seconds and rip him to shreds before he could even get off his second ball. He searched for another fifteen minutes, finding nothing. Satisfied that it was a clean miss, he relaxed and checked his powder.

He thought about the men he had known in the Expedition and the Indians he had traded with and fought alongside in battle. Many of them were dead. Coming into the country and trying to make a living, trying to open trade, was it all worth it? Was the price of progress worth it? Could what he and the Expedition accomplished even be called progress?

The Crow seemed to be better off for it. The new trade possibilities benefited them. He liked the Crow. They were honest traders, reliable warriors in battle, and impeccable horsemen. They were well-built with good hygiene, had good teeth, and were virtually lice-free. But what he liked most about them was that they did not live for war. They preferred peaceful trade and commerce to constant warfare. Of course, they were renowned horse thieves; their young men had to prove themselves one way or another. This is what the Expedition never understood, he thought. Total peace could never be realized.

He continued downriver. The wind picked up, and the cold was numbing his bones. After another hour, he pulled ashore and decided to eat and look for fuel. He found some dry wood in the brush along the riverbank. It was enough for a fire or two, but he thought it would be wise to gather extra and carry it with him since he did not know where, or even if, he would make camp that night. He climbed up the bank, his moccasins slipping slightly in the sand, and found a worn path, likely a deer trail. There was a small stand of dead cottonwoods that had blown over, and he went to work with his hatchet cutting off branches. He worked methodically, stopping ever so often to scan the woods around him and peer at the river for any sign of a party. He always found it strange that a man could travel for weeks without coming across another person, or even any sign of one, and then could suddenly encounter a hunting party, warriors, other trappers—anyone—and be killed in an instant.

He continued to work and listen intently for any hostile sound. After an hour, he stopped, stretched his back, and looked up at the cloud cover hoping for some respite from the cold. Then he saw movement out of the corner of his eye and lunged for his rifle. He brought it to his shoulder and found the trigger. He caught movement again and realized that whatever it was, it was small and four-legged. His stomach rumbled as the animal came into view. It appeared to be a coyote, but it was steadily coming towards him even though it could clearly see him standing there in the trail. As it came closer, he saw that it was a dog, quite tame, and probably not far from its master. The hair stood up on his neck, and he checked his powder. He thought about shooting the dog for meat, quickly gathering his wood, and paddling as if his life depended on it. The shot would be heard for a mile, at least. He stared at the dog while it sniffed him and decided to head in the direction it had come. He never liked the taste of dog, and if a party was close, he wanted to see them first. He threw a hunk of jerked venison into the woods to keep the dog occupied and slowly worked his way up the trail with his rifle leveled. The trail rose steadily, and he found himself walking uphill for several hundred yards until the trees began to thin. There was a moss-covered clearing and rock-outcropping ahead. He couldn't quite see, but he assumed this led to a cliff or overhang. He wanted very badly to look over the cliff into the river below, a section he would soon be paddling through. He thought the risk of exposing himself was worth discovering who or what might be there. Once he exited the trees, he began to crawl on his belly, inching towards the edge of the cliff. He stopped often and listened, but he could hear nothing except for the soft cadence of the river. He eventually came to the edge and peered over. Down below, there was an Indian squatting and appearing to mend something. He watched intently, unsure if it was a lone traveler or just one member of a larger party. The Indian stood, disrobed, and jumped into the river. He could hear gasping, even from the cliff, and understood the feeling; he had often bathed in icy rivers himself. The Indian rose and waded back out of the river, having only been in the water for five seconds or so. Now that he caught a frontal view of the naked body, he saw, quite plainly, that it was a woman. He saw her breasts bouncing as she hopped onto the sand, and he saw the dark patch of her genitals. Her face was hidden as she used a blanket to dry her hair. A very intense feeling of desire took hold of him, and he watched transfixed as she dried herself, put her clothes back on, and squatted beside the fire rubbing her body and singing softly to herself.

He thought about what he had given up. Many nights he lay awake in the woods and thought of his friends who had left the mountains, married, and had a family. He felt such intense envy that they had a woman next to them every night, whereas he had a woman, at most, a handful of times per year. Often, he would fall asleep beside the dying fire and dream of naked women coming out of the woods, so many he couldn't even count them. They would sit around his fire and stare at him, smile, but say nothing. And then he would wake up.

He thought quite seriously about descending the cliff, having his way with the woman, and leaving. Taking her with him was also an option, although risky and cumbersome and would greatly increase the chances of being killed. He surveyed the area and tried to calculate the odds that she truly was alone, save for the dog. If he descended, he would most likely have to kill her. He had never killed a woman before, but many of his friends had and felt no compunction in it. He watched as she sang softly and rocked herself back and forth. He tried to imagine why she would be out here alone with just her dog. Was it even her dog? Was there another party close? Was she an outcast of some kind—a witch?

She removed the cloth that had been covering her head to facilitate the drying. As her face came into view, he felt a sudden and intense nausea. Even from that distance, he could make out the signs of smallpox disfigurement. The blisters and sores covered her entire face, forehead, and neck. He figured she was probably a Ree woman since he had heard recently that a smallpox epidemic was decimating the tribe. He watched her sitting by the fire. He decided that he would leave her for now, but if she was still there in the morning when he passed through, he would have to kill her. Doing so wouldn't be so terrible a sin, he thought. He retreated the mile or so back to his canoe by the river. The sun was beginning its descent to the horizon, and he thought, given the circumstances, that he should make camp and leave in the morning.

He gathered a dozen stones and formed them into a circle. Then he placed several of the cottonwood branches he had cut in the center of the stones. He retrieved a bundle of cedar shavings that he carried with him, separated a small section, and placed it below the smallest of the cottonwood branches. He poured a very small amount of powder onto the bundle and brought forth his flint and steel. It took on the second strike, a record for him, and he cupped it with his hands. The ember glowed as he softly blew on it, and smoke began to bellow out. He sat back just a bit to give the nascent fire a chance to catch on the cottonwood. He didn't need the fire for cooking. He still had some jerked venison that he had brought, although the supply was running low. He knew he would have to do some hunting sooner or later but hoped to reach the fork and the traps before that. He didn't like the idea of wandering around hunting and stumbling into a party before he had gotten to the traps. Once he was headed home, he might have to take some chances.

The fire was crackling at this point—the cottonwood had caught—and he could feel the first warmth of the coals against his body. He inevitably felt anxiety as he watched the glow. Every night he camped with a fire increased the risk of torture and death. When he first came into the country years ago, he rarely used a campfire. He was much more paranoid then because of his inexperience. Now he liked his comfort more and often decided to take the risk, even though the threat had probably increased over the last five or six years.

As the fire burned down, he thought about why he was on this journey. Why was he risking his life for some rusty traps that very well may be beyond use? Why did he even live this life in the first place? Why couldn't he settle down with a wife and children, maybe farm some, and put this life behind him?

When he wasn't in the country, when he was at fort or in Saint Louis, he felt mildly ill, not physically, but as if he wasn't fully himself, fully alive, just a kind of skeleton that went about his business with no zest or intention, like a lame horse in a corral. Perhaps, he thought, this feeling would fade as he aged. He might settle down eventually. He longed for that. He longed for a time when mountains and rivers and beavers and traps no longer held a spell over him. A time when he could live a normal life and be happy with it and leave this Godforsaken country behind him.

He struck camp at first light. He threw sand on the fire and urinated on it. Then he repeated the process several times until there was no smoke. He loaded his supplies into the canoe, checked his powder, and headed into the river. He paddled slowly and deliberately, listening intently for any sound or disturbance. He knew the Ree woman's camp was a little over a mile ahead of him, and he wanted to be prepared. He thought again about what he would do if he came upon her. He decided that killing her was not necessary. He would sign to her as he paddled by, and if she made any hostile movements, he would shoot her. He hoped she would watch him passively, and he could move on without incident.

He saw the rock outcropping ahead where he had lain the day before. It was behind a bend in the river so he could not see below it just yet. He held his paddle with his left hand and the rifle with his right and leveled it ahead of him. He listened for the sound of singing. As the river bent towards the cliff, he looked along the bank. The woman was lying there on the sand, completely naked. Her eyes were wide open, and there was dried blood on her face and neck. She had been scalped. He slowed his canoe and landed about fifty yards upriver. He found the dog there as well. It had been skinned, boiled, and eaten, just the skull and some bones remaining. The camp was torn apart, anything of value taken. He wondered why her assailants had fallen upon her, diseased and highly contagious as she was. She probably camped without a fire, and they came in the night and hadn't noticed her condition until after they had raped and killed her. Or maybe they had smallpox themselves. Or maybe they just didn't care. He figured it was a Blackfoot war party. Now he had to wonder whether they had come from above her on the river or below. If they had come from above, they would have passed by his camp and surely attacked him. Perhaps his fire had died out, and they didn't see him. That seemed improbable, and he deduced that they came downstream from her, killed her, and for whatever reason, did not come further upriver to his camp. This, of course, meant they were likely still in the area, maybe even watching him right now.

He stared at the woman and thought of all the

people he had seen dead or dying—men, women, and children, often butchered beyond recognition. She had smallpox, but that wasn't enough. She had to suffer even more. He looked at the remains of the dog—a dog that he had met the day before and had gladly eaten venison from his hand. One man provided food, and another butchered him and ate him. More and more, this country was exacting more from him than he could give.

He shoved off and headed downriver. If the Blackfoot were still in the area, they might see him. But if he stayed until nightfall, he would lose too much valuable time. His supplies were running low, and the traps were rusting in the river. He wondered how long it would go on. Trade, raiding, disease, deaths. It was a cycle that could not be broken. The Expedition wanted to bring peace to the tribes. They would pacify peoples who had been fighting each other long before any European came into the country. They called it progress and said it was best for the natives, and that they would benefit in the long run. The people that benefitted most, he thought, never fought in a battle, built a fire at twenty below, or watched while their friends were tortured and killed.

He was told it was a noble calling, but the carrying out had proven to be bloodier and more costly than he could have ever imagined. Pacifying peoples so trade could flourish, land could be settled, and resources harvested sounded good in meetings and letters. The reality ended up being dead women and boiled dogs, and a man risking his life for some rusty metal to catch fur for someone in another country to wear and never have to think about where it came from, or how it got there, or what it meant.

He made a pact with himself at that moment. If he retrieved the traps, he would return to the fort, sell everything he owned, and save his rifle and hunting knife. Then he would find someone to marry, buy a plot of land, and try his hand at farming. Six years ago, the thought would have been ridiculous. Now, he decided he'd had enough of the mountains and the wilderness and the killing and the violence. Growing a family and a crop would be his second act. He would remember this trip and the woman and the dog. He would remember his time in the Expedition and the beaver trade. He would think about the skill and luck that had seen him through when so many others had died, and he would try to use his knowledge and experience and apply it to building something. And maybe his time on the line wouldn't have been wasted.

He paddled on and prayed for the first time in his life. Then he looked at the stretch of river ahead of him and felt the cold numbing him in the canoe. He checked his powder and adjusted the hunting knife on his belt. He smiled at the absurdity. After the life that he had lived, what god would listen to him now?

11

CITIZEN HELPLESS

We are loath to let others do unto us what we happily do to ourselves. —**Chauncey Starr**

John Colter made his way into the history books of America largely because he was an original member of the Lewis and Clark Expedition. For the most part, he proved to be reliable, competent, and dedicated. When the Expedition ended, Colter returned to the wilderness to trap beaver in the upper Missouri River valley. In addition, he searched for and found the sources of the Yellowstone and Bighorn Rivers. Today, he is credited with discovering the country that eventually became Yellowstone National Park. Much of Colter's experience is unknown because he left behind no written record. Even his legendary dash known as "Colter's Run," where he outran and hid, naked and weaponless, from hundreds of Blackfoot warriors intent on killing him, was relayed by Colter himself through word of mouth.

Colter embodied an archetype of the American experience in the early 19th century. Bored and frustrated with life in forts, he set out into the wilderness, sometimes on his own, sometimes with partners, and attempted to make a living trapping beaver. His travels brought him into contact with many Native American tribes, most notably the Crow. Colter befriended the Crow, traded with them, and fought alongside them. In the summer of 1808, he was traveling with a large party of Flathead and Crow in present-day Montana when a Blackfoot army fell upon them. Colter was wounded in the leg and incapacitated during the battle, but he survived, and they defeated the Blackfoot despite being heavily outnumbered.[1]

Many men of Colter's era trekked into the wilderness to eke out a living trapping, trading, and whether they wanted to or not, fighting. The mountain man lifestyle was extremely dangerous, unforgiving, thankless, and impoverished. Most, including Colter, ultimately gave it up for a more settled existence, relatively speaking. After years of such a spartan and precarious existence, the comforts of marriage, children, and home must have become increasingly attractive.

Colter's lived experience demonstrates an American exerting maximum agency in a time when that was relatively commonplace and far from exceptional. But his life was hard, dirty, and violent. It's not 1808 anymore, and the world is a very different place than the one Colter and his ilk knew. And we should be thankful for that. We are the lucky successors to the sturdy and resilient men and women who settled this country, molded it into the relatively well-functioning society we have today (often against their will, as our nation's legacy of slavery can attest), and provided most of us with a higher standard of living than the one they had.

The problem that has arisen is one that would have been very foreign to John Colter and others living in 1808. It would have been foreign to a citizenry living in a society where doing for yourself was an ideal. And those who did, like Colter, are enshrined in our nation's lore as examples of grandeur—the ultimate rugged individualist. The issue we have encountered is that citizens in our modern society who dare to exert agency over their own lives, as Colter did, are increasingly chastised, shunned, and vilified. We are now living in a country where procuring your own needs through your own efforts, is becoming anathema to what our technocentric culture is proselytizing. Technology has become the ubiquitous and central force in our lives and has increasingly normalized behaviors that would have gotten John Colter and others like him killed. What in God's name are you talking about? Bear with me.

For the purposes of this book, the term **"Tech"** will be defined as the following: Tech consists of non-government, large-scale corporate entities that operate in the digital, online, or remote atmosphere, i.e., the cloud, and whose fundamental goal is to eliminate human agency in order to facilitate comprehensive dependency, exploitation, and sustained financial profit. This implies, but is not exclusive to, the possibility that said entities are working together in a coordinated campaign. Facebook, Twitter, and Amazon are prominent examples.

Another term that will be prevalent in this book is **"synthetic protein."** This term will be defined as any protein that has been manufactured or grown in a lab, and/or did not come directly from a slaughtered animal that was raised on a farm or lived in the woods.

No matter what time in history or in what location they were born, every person grows up in a society that holds certain cultural standards as normative and incentivizes and disincentivizes behaviors based on this paradigm. Every culture is different. We think nothing of giving women in the US the right to vote, go to school, work, drive, and wear provocative clothing, if they so desire. In other nations, these constitute crimes. But on a grander scale, the overriding framework that has institutionalized normative behaviors and greatly influenced how our citizens interact with each other, more than any other, because of the prominent Judeo-Christian tradition that is prevalent in American culture, is the Ten Commandments. This is the dogma through which every child in our country becomes indoctrinated through school, worship, and media. And this isn't necessarily a bad thing. We could do a lot worse than "Honor your mother and father" or "Do not steal." But it gets a little murkier when we get to "Thou shall not kill." If we look closely at American history, the commandment should really read, "Thou shall not kill except in very specific instances sanctioned, encouraged, and often mandated by the federal government." As Jared Diamond writes in *The World Until Yesterday: What Can We Learn From Traditional Societies?*:

> *"In Western state societies today, we grow up learning a universal code of morality that is promulgated every week in our houses of worship, and codified in our laws. The sixth commandment declares simply, 'Thou shall not kill'—with no distinction between how we should behave towards citizens of our own state and towards citizens of other states. Then, after at least 18 years of such moral training, we take young adults, train them to be soldiers, give them guns, and command that they should now forget all of that former upbringing forbidding them to kill.²"*

The Vietnam War comes to mind. With Vietnam, there was no Pearl Harbor or 9/11. There was no catastrophic attack that killed thousands of Americans; it was not a war of self-defense. We went to Vietnam for purely political reasons. Therefore, we needed an unwritten social doctrine to incentivize an eighteen-year-old man to go to a foreign country and kill people he has never met. This may not square with the Ten Commandments that he had been taught in school, but it is OK because the government has legalized, even mandated (through the draft), his participation in another nation's civil war. This is what's best for him, what's best for the country, and therefore, he should capitulate. But it is *not* OK or sanctioned by the government for him to kill someone domestically for political reasons. We certainly cannot tolerate Democrats and Republicans fighting in the streets. That is not a recipe for a stable society. These are the distinctions that must be made, and the citizenry should not question these inconsistencies because the government knows what is best for them, and they should be grateful.

These societal and cultural norms can shift over time. This unwritten social doctrine has been challenged. Increasingly, the majority of the American people and many of our elected officials would not be in favor of sending young men and women into battle in another nation with whom we are having a political dispute. It would be very difficult for an American president today to insert regular ground troops into a foreign civil war, such as Vietnam, if that war posed no potential or tangible threat to our national security or the well-being of the American public because of the shift in what has become, to the citizenry, acceptable reasons for waging war. At least he or she could not do so if he or she needed to win re-election. I hope this is one of the lessons we learned from Vietnam.

There are many examples throughout history of governments incentivizing certain behaviors and disincentivizing others in order to achieve the government's goals. They have enacted societal and cultural change, sometimes slowly, sometimes quickly, amongst their citizens. A classic example of this is Nazi Germany. Upon coming to power, Adolf Hitler convinced many Germans, who did not previously hold this belief, that Jews, homosexuals, gypsies, and the mentally and physically disabled, among others, were responsible for Germany's economic, social, and military weaknesses. And therefore, these groups had to be eradicated, requiring a nation-wide racial cleansing. Furthermore, Germany had to acquire more land space, annex Austria, invade Poland, and initiate a war that would ultimately kill seventy million people. Governments are capable of shifting norms to achieve their goals because, in many instances, the citizens have given them the power to do so.

What about non-government actors? What about third-party entities that have not been elected by the citizens and have not, ostensibly, been given specific powers? How are they to achieve their goals? How can they incentivize and disincentivize certain behaviors amongst the masses? What tools do they have at their disposal? And do they really need to jail and kill people who might disagree, such as what took place in Germany, Russia, and Italy during World War II? Those antiquated methods are too difficult, bloody, and short-lived. Ultimately, such nefarious goals are unable to achieve long-lasting success in some cases because of outside forces (the Allies capturing Berlin) or in other cases because the bulk of the citizenry will no longer stand for it; there are revolutions, or the government leader is removed through elections, coups, or even assassinations. Things did not end well for Hitler or Mussolini. Non-government actors have learned through the trials and errors of governments.

Ultimately, for those modern citizens who refuse to capitulate, Tech will incentivize citizens already indoctrinated into conducting intimidation campaigns. Tech has nuanced methods to convince its citizens that it really does hold their long-term interests in the highest regard. And it will do anything to increase and sustain their happiness, well-being, and fulfillment. Tech has systematically provided products and devices that make our lives "easier" and less labor-intensive. We have smart automobiles, smart televisions, smart thermostats, smart refrigerators, smart washers and driers, smart radios (Alexa), I-Phones, Bluetooth, social media, online dating sites, streaming services, data backups in the cloud, online marketplaces (Amazon), and digital applications for virtually every single human need or desire. If an application doesn't currently exist for a specific desire, it will have been created by the time I finish writing this sentence. Some of these devices are currently completely controlled by Tech. Eventually, all of them will be. Our homes will warm and cool, our cars will start and stop, our phones will function or malfunction, all based on the desires of Tech and the degree to which we have, individually, embraced and complied with Tech to its satisfaction. As for those citizens who reject this culture change, who refuse to accept what Tech is offering, who would still rather hunt, fish, farm, and garden for their food, or refuse to buy a cell phone, or insist on rising from their recliner, walking over to the thermostat and manually (it really is barbaric when you think about it) turn the dial? Well, they have to be convinced.

***

In November 2018, I killed a deer with a rifle during
Missouri's fall deer season. It had snowed heavily that
morning, six or seven inches, which was exceedingly rare
for November. A medium-sized buck crossed my face
at 8 am on the dot while I was sitting in my tree stand.
He was walking lazily along the fence line, stopping ever
few yards to survey the area. He eventually jumped the
fence and turned broadside in front of me. I slowly raised
the rifle to my shoulder. In doing so, the stock rubbed
against a zipper on the collar of my jacket. The noise
was just enough in the stillness of the freezing morning
air to stop the buck, suddenly, forty yards away. He
knew something wasn't right. I fired a round through
his lungs before he ever figured out what it was. It was
the closest thing to a "chip shot" I have ever had while
hunting. He ran fifty yards or so, downhill, and piled up
at the bottom of a ravine. He died within thirty seconds
of the bullet passing through his body. My satisfaction at
having dispatched him so quickly and cleanly was short-
lived. I realized that I was now going to have to drag one
hundred fifty pounds of dead weight uphill, in the snow,
several hundred yards to a place where I could haul him
out with a neighbor's tractor and back to the house for
processing. I made my way down the ravine, grabbed
him by the antlers, and set them on my thighs. Then I
began to pull. I probably needed the workout anyway.

It took about four hours to haul the deer back to
my house, hang him on a gambrel, remove the organs,
skin the carcass, and then butcher and wrap the meat.
And I still had a couple more hours of work beyond

this in grinding various cuts into burger. All in all, it was probably five to six hours total if one did it straight through, from when I pulled the trigger in the woods to when I was finally cooking ground venison in my skillet for dinner. It was a very labor-intensive activity as anyone who has done so can attest. But the yield was fifty pounds of venison that I didn't have to pay for, didn't push a button to get, that wasn't sent to a fulfillment center, didn't come in a box, or have a tracking number. I didn't have to wait for Tech to feed me.

Any citizen who is willing to satisfy some or all of their needs is an enemy of Tech. Tech cannot stand for this. One citizen with agency leads to two, then to four, and then the house of cards that Tech has built begins to fall. So how will Tech convince wayward citizens like me to cease doing for themselves and procuring their own food? The first step is to bombard citizens through every channel available, promoting how easy, clean, and cheap life is once one has embraced Tech. *Don't hunt, fish, farm, and garden for your own food! No! You don't have to do that! Why would you want to? It's hard! It's time-consuming! It's dirty! It's dangerous! It's violent! It's...wrong! No, no, we have that problem solved for you! You need protein? No problem! We are going to make synthetic protein for you in a laboratory! Synthetic beef, synthetic chicken, synthetic pork, synthetic milk, synthetic eggs, synthetic turkey, synthetic venison, synthetic lamb! All you have to do is go on Amazon. com and push a button! It will arrive a day later at your doorstep in a climate-controlled box, at a low cost, with minimal impact on the environment! This is progress!* And we will all talk about how great Tech is, and how easy life is now that we have fully embraced it. Why didn't we think of this sooner? Why are we working so hard? Think about all the free time we now have to spend on social media. How did we ever survive before Amazon?

Progress always comes with a price. And often, it is paid by someone or something other than what was intended. When people no longer have to hunt, fish, farm, or garden for their own food, when Tech has solved that "problem" for them, the statement "you don't have to do anything about it" becomes something different. Once Tech succeeds in convincing the federal, state, and local governments to outlaw hunting, fishing, and animal agriculture for individual citizens, that statement becomes something quite different altogether. It is no longer, "*you **don't have to** do anything about it.*" It becomes, "*you **can't** do anything about it.*" The price of our relentless obsession with "progress," our relentless obsession with automation and technology, is human agency. We are adopting a learned helplessness, and our appetite for dependency is growing. Tech will continue to work tirelessly to become our only source for every need and desire. This is what's best for us, and we should be grateful.

I used to own a landscaping business in Saint Louis. I started it from scratch, ran it for fifteen years, and eventually sold it. I was providing a service for interested customers. For those individuals with some disposable income who would rather not spend their weekends toiling in the brutal Missouri heat and humidity, I offered an option for them. However, I was not attempting to create an environment or a world in which people had no choice but to patronize my business, no other options, and therefore had to take what I was offering. I did not make incessant phone calls or send threatening emails and texts to those citizens who still chose to mow their own lawns, rake their own leaves, and shovel their own snow. I did not question their reasons for doing so, did not attempt to intimidate them into cease doing for themselves. It would have been offensive, mean-spirited, poor business tactics, and simply amoral. I likely would have been prosecuted for harassment and reported to the Better Business Bureau. My customer base would have dwindled, and before long, I would have been out of a job. There is a difference between offering a service to a customer and completely eliminating a person's agency to the point where they have no choice but to accept my "help." Why are we allowing Tech to do to us what would be wholly unethical and criminal behavior in any other arena? What power does Tech hold over us, and why do we increasingly feel that it is the most necessary and intrinsic component of our daily lives?

The greatest threat to Tech is a well-informed citizenry with agency. Tech's slogan is: ***Do not do for yourself what Tech can do for you better, cheaper, cleaner, and easier.*** If you are hungry, push the button. Thirsty? Push the button. Cold? Push the button. Hot? Push the button. Bored? Push the button. Stressed? Push the button. We have become a Tech-curated, push-button society. If you refuse to accept this culture change and continue to do for yourself, you will face certain consequences on a sliding scale relative to the offense and the degree to which you publicize your participation in that offense to other citizens. In the world Tech has created, perception truly is reality. Tech knows where I live, where I work, what I drive, who my family and friends are, who my neighbors are, my pet's names, my age, my height, my weight, my birthday, my political leanings, my financial status, my sexual orientation, my marital status, my health status, my IQ level, my favorite book, TV show, and movie, my musical preferences, my travel habits, my criminal history and more. Tech will make all of this information available to its most loyal and indoctrinated citizens. Let me provide some clarification on this point:

Several years ago, a young girl in Texas—maybe thirteen or fourteen years old, I can't remember exactly, raised a show steer as part of the 4-H program in which she was a member. She posted photos online of herself feeding the steer, grooming the steer, caring for the steer, and showing the steer. Ultimately, she publicized the last time she saw the steer—the day it was sent to the slaughterhouse. The outrage online was immediate and horrifying. She and her parents were vilified. The girl was called names that would make George Patton blush so you can use your imagination. It was shameful but very predictable.

Should we be surprised? At the end of the day, what was this girl and her family doing? They were raising an animal for food independent of Tech. They dared to do for themselves what Tech wanted to do for them, and what was worst of all, they were publicizing this on social media. Tech cannot stand for this. Tech needs this girl to need it. Now, other parents might see this, might think it is a good thing, might see what this girl is learning about small-scale animal agriculture, responsibility, and independence, and might even consider enrolling their own children in such programs. This child was dangerous. Tech has convinced its most fervent followers that it is waging a war, that they are fighting a battle of good vs. evil. This is how previously stable citizens can and will justify to themselves, and others, that threatening a child is acceptable because she raised an animal for food.

Tech's reaction was so harsh because it has been trying desperately to shift our societal and cultural norms. What has changed in our culture in the last twenty years? Who is engineering this change, and most importantly, who stands to benefit from it? Tech's most fanatical operatives are often people you would never expect. Karl Marx wrote, "Religion is the opium of the people." Well, Marx never heard of Twitter. He never heard of Facebook, TikTok, Snapchat, or Instagram. To him, Amazon was a rainforest, and Google was a number. Marx never heard of the internet or social media. If he had, he might have written something different. In our modern age, even the most sane and self-aware citizen can become radicalized very quickly. Our digital opium has become more addictive than any religion Karl Marx could have imagined.

This young girl in Texas was reinforcing her own agency. She was learning to become an independent and capable citizen. This cannot be tolerated. Perhaps, she and her family could not be intimidated to cease and desist. But the battle is not going to be won or lost with citizens such as her or myself, the outliers. The battle will be won or lost with the masses. If those parents and children who were even remotely considering becoming involved in animal agriculture saw the threats and intimidation flying down-range at this girl, they might think twice. They might think to themselves, "I don't want my child to be called bad names. I don't want my child to have low self-esteem or to feel ashamed or guilty about participating in an activity that we hoped would build character and promote independence. Maybe I will teach my child to play golf instead and leave food production to someone else (i.e., Tech)." This is the desired outcome, and Tech will have won.

Tech will win by creating and sustaining dependencies amongst its citizens. A citizen with agency is much more difficult to control. Therefore, Tech must do everything in its power to render us helpless. Because a helpless citizen is vulnerable. A vulnerable citizen can be made desperate very easily. And desperate citizens will do what they are told to do. They will eat what they are told to eat, buy what they are told to buy, read what they are told to read, and believe what they are told to believe. Why? Because they have to. Now they have no choice. Tech has methodically and systemically convinced them to surrender their agency a little at a time until there is no more left to give.

Eventually, every action a citizen takes will be monitored and recorded by Tech to track compliance or non-compliance. Even those recalcitrant individuals who stubbornly cling to the last bastion of freedom will be monitored, recorded, and punished, if need be. If I choose to walk out of my house at 3 am and go into the woods to defecate, Tech will know about it. There will be a Tech-controlled drone hovering over me with a spotlight recording and measuring every metric of my experience. *What color was the waste? What was the consistency? What was the overall satisfaction level? Did the citizen feel like the process took too long or was too short? Did the citizen need five squares of paper this time or four squares like his previous three shits?* The drone will then collect and analyze a sample of my waste to determine if it contains any non-Tech-produced and distributed food. *If so, where did it come from? How did it get into his body? What should be his penalty for non-compliance? What Tech-controlled devices in his home can be disabled until he exhibits proper behavior? What Tech-sponsored products can be advertised during his experience? Could he be sold a better brand of toilet paper to reduce the total amount of squares used? Can he be sold a more potent laxative to facilitate better passage? What should be displayed on the drone's screen based on his user history? What music should the drone play to better enhance and curate his experience?* But the most critical question of all that Tech will ask is: *Why in hell is this citizen shitting in the woods at 3 am when he could be in his house on his sensor-warmed porcelain smart toilet where it would be much easier to monitor him and his experience and better cater to his needs if he would just cooperate? What is wrong with him?!!! Who shits in the woods at 3 am?!!! After all, we already blasted over all Tech-sponsored channels that citizens can defecate on their smart toilets while they post to Facebook and scroll through their Twitter feed! As soon as they think about shitting, the toilet seat starts to warm!!! This is progress!!!!!! Why has this citizen gone rogue?!!! Is there any chance, even a small one, that more citizens will begin shitting in the woods at 3 am? What else could this lead to? This could really be a problem! We have to prevent this!!! How do we punish this ungrateful Luddite?!!! Wait a minute. Wait a minute. Wait just a Goddamned minute!!! Isn't this the guy who murders animals in cold blood to compensate for his mental and physical shortcomings and encourages others to do likewise? Let's pour that over our channels with the appropriate information and direction for our citizens. Problem solved.*

III

JUST A COW?

I seem to have been born with an aptitude for a way of life that was doomed. —**Wendell Berry**

I used to raise cattle full-time on my farm in Missouri. I really enjoyed it, but I learned the hard way that you might not starve on a farm, but you will go broke. And that's what happened to me. I eventually had to get a day job driving an ice cream truck, and now I raise cattle and chickens part-time. I am not going to try to convince you that cattle are the most intelligent animals on the planet in this chapter. They aren't. However, they aren't as dumb as many people think. And one can learn a lot from working with cattle, if one pays attention. And I have learned a lot. I have learned a lot about herd mentality, leaders and followers, and the benefits of going rogue.

On a small farm like mine, working cattle is a fairly simple and straight-forward endeavor. When you first get into the business, you will probably go to the sale barn and buy cow/calf pairs to start your herd. You will transport them to your farm, unload them, and give them a few days to get used to the place. You will check on them every so often and watch for any indication of problematic or undesirable behaviors. After the cattle are used to seeing you every day, out in the pasture with them, they will get accustomed to you, maybe even start to like you. They begin to lose all fear of you. To facilitate this loss of fear, you can bring out a bucket of feed, sweet corn, range cubes, or molasses. You will pour it on the ground, and they will eat it. After a while, they will start walking towards you as soon as they see you with the bucket, even before you have poured it on the ground. It isn't long after that before they are running towards you. You want your cattle to react without thinking. I can tell you from experience that the ones that think for themselves end up being nothing but problems.

You will want to identify your Lead Cow as soon as possible. This is usually an older cow that has been in the herd for a while and has, through a pecking order of sorts, risen to a leadership position. When she decides to eat, the others will eat. When she goes to water, the others will go to water. If she heads for the feed bucket, the others will follow. Your Lead Cow is vital to the successful management of your herd.

There are certain times throughout the year when you will need to bring the herd into the corral for various reasons. They may need vaccinations, deworming, pregnancy checking, castration, etc. And then they will be released back into the pasture when the work is done. There are also times, however, when individual cattle will not be released. They will be loaded onto a trailer and shipped off the farm for an eventual rendezvous with the processor. I sell most of my beef direct to the consumer, so when finished steers leave my farm they head directly to the slaughterhouse, and the lights will be turned off in 24-48 hours. This is the reality of food production Tech is trying to convince us is wrong. ***Killing animals for food is abhorrent, savage, and barbaric. We have moved past that stage in our evolution, good citizens. Synthetic protein is the way of the future. Now sit in front of ~~our~~ your screen, push the button, and wait for it to come to you.***

Cows, like people, have personalities. They may not be as complex, but they have them. You have your friendly cows, your ornery cows, your "dumb" cows, your "smart" cows, your flighty cows, your laid-back cows, your goofy cows, your mischievous cows, your moody cows, and your rogue cows. The last category is a rare occurrence if you choose your cows wisely and are selecting for certain traits within your breeding program. But you do come across one or two from time to time. This is the cow that, despite your best efforts, never fully trusts you, never buys in wholesale to the program. She may eat the sweet feed you pour on the ground. She may cooperate when you move the herd, but she will only let you get so close to her. She is always the last one to come into the corral, if she comes in at all. No matter what you try, she just won't accept you. You can see it in her eyes.

Well, this is a problem. This makes your job harder, more time-consuming. You can't have this. One wild cow can get the others riled up—her disobedience can spread like a disease throughout the herd. Where you initially had a docile and agreeable herd (an objective that was not achieved without significant time and effort), you can, over time, find yourself with one that is unmanageable and unprofitable. Often, the best thing to do with a rogue cow is to remove her. In my experience, the wild ones stay wild no matter what you try, no matter how kind you are. They are within the herd, but not of the herd.

Of course, "removing her from the herd" is a lot easier said than done. You will have to bring all of them into the corral and try to separate her to be shipped out. She might realize what is happening and completely lose her mind. She will ram the corral panels and attempt to jump over them. She may put her head underneath the bottom rung and fold the panel in half with a sharp upward motion, or just start bucking like a rank bull in the chute. She may kick other cows, calves, even you. If you've been through this before, you tend to think it isn't worth it. Maybe it's better just to leave her out on the pasture to her own devices. You could shoot her, sure, and eliminate her that way, but that might instill fear of you in the rest of the herd who will witness it, not to mention cost you a good deal of money. You could tranquilize her and remove her, but that's very expensive and time-consuming as well. Maybe it's best just to leave her be. She may let the bull breed her, and you might get a calf out of her that you can sell quickly, even before weaning, when it will have picked up a lot of its mother's temperament, and at least get some money out of her that way. It may not be the end of the world. After all, one rogue cow isn't worth losing sleep over.

The main issue arises if this cow is able to spread her discontent to the rest of the herd. This must be avoided at all costs. You will pull every card you have to prevent this, every single one. And, if you are a halfway decent cattleman, you will already have an Ace-of-Spades up your sleeve, ready to be played. A card you have been nurturing and grooming and incentivizing for some time—your Lead Cow. If you have properly facilitated her status within the herd as well as her loyalty to you, she will prove invaluable. She will do more to condition the herd's mentality to your liking than anything else. Over time, the rest of the cows will feel that their best interests lie with her (you) and will act accordingly. That wild and disobedient cow will be treated as an outcast, an undesirable, and her status within the herd will reflect this. With any luck, she will have no tangible effect on the herd's temperament, just a minor inconvenience in the end.

Tech has also groomed and conditioned its "Lead Cows" to properly motivate the rest of the herd. Tech calls them "Social Media Influencers." I call them Lead Cows. We receive constant direction and guidance from these influencers and are led to believe we can only live our best lives if we consume their messages as fast as possible, swallowing everything whole like so many pounds of corn and molasses. These influencers are held up as examples of prosperity in the world Tech has created, despite the fact that many of them have done nothing to earn that status—at least in our eyes. But in the eyes of Tech, their loyal, sustained, and predictable allegiance to any and all Tech-sponsored platforms, products, and publicity has made them invaluable—the steady hand when the occasional rogue cow garners attention and riles the herd.

These Lead Cows pervade every aspect of our daily lives. Through their messaging, we learn that we must be smarter, richer, better looking, healthier, more fashionable, more "modern" (read digitally connected). We are supposed to want to be the influencers. And we can only achieve this goal by purchasing Tech-sponsored products and, most importantly, spending as much time as possible engaging with Tech on its platforms and on its terms. Only then will we reap the rewards of such loyalty: wealth, status, fame, and happiness. Anyone who does not follow their example will end up alienated, impoverished, and miserable. *And worse, your children might end up this way. They might grow to middle age with little to show for it socially, financially, and occupationally. They might become isolated, disturbed, and even engage in criminal activity. And horror of all horrors, they might end up living alone, in the middle of nowhere, pushing ice cream on the street for a living and murdering animals—a paranoid writing a psychotic manifesto railing against something that doesn't exist!!! Do you really want your child to become such a person?!!! Look at him for Christ's sake!!!! No, I didn't think so. Now, be a good obedient cow and follow your Lead, learn from her, and emulate her. And when you see the feed bucket, run as fast as you can!"* And so we will. We will follow the influencers and heed their advice to the best of our ability. We will place complete trust in them because they know what is best for us. We will never even look to see who is holding the bucket or question their motives or what is in store for us. Once we have become addicted to the feed, we will choose it over our freedom every time. We will have been thoroughly conditioned by this point. We will gulp down the feed faster and faster until we can hardly breathe, and then we will gallop full speed to the slaughterhouse like so many fat steers smiling the whole way.

The Rogue Cow will hang back along the trees and watch this spectacle with tired eyes. She will look on with a knowing and wearisome indifference as if she had been through it all before. After the trucks and trailers have left and the herd is gone, she will calmly turn her back, walk into the woods, and spend her remaining days alone, hiding out like a criminal on the run. She will wander from farm to farm, living off the land as best she can. At night she will venture out to graze in the fields and drink from the creeks and ponds. It will be a lonely and difficult life, but a life lived on her terms. Eventually, she will grow old and decrepit. She will have trouble walking, become partially blind, and lose most of her teeth. And one day she will lie down and die. She will bloat, rot, and fester, and the coyotes will gorge themselves on her carcass until there is no meat left.

Then they will move on to their next meal and never look back. The grass will grow up, covering her hide and bones in the woods, and in time, it will be as if she never existed at all.

IV

WORK SETS YOU FREE

You are fools to make yourselves slaves to a piece of fat bacon, some hardtack, and a little sugar and coffee. —**Sitting Bull**

Tech has embarked on a systematic and sustained campaign to convince all of us that work is no longer necessary. Only unenlightened and ignorant people will work in the future. The rest of us will be freed from this banality of daily existence—free to write, draw, sing, and create. Life will be much more enjoyable because we will be able to spend all our time on meaningful projects, hobbies, and relationships. Many of us have marveled at how hard our parents and grandparents worked and sacrificed so we could have the lives we have today. Future generations will marvel that people worked at all. Tech will mow your lawn, haul your trash, fetch your groceries, cook your food, clean your home, tie your shoes, brush your teeth, and bathe you. You will sit behind a screen for recreation, entertainment, worship, and more. You will have complete and total automation in every aspect of your life. It will be a beautiful world to be sure. Why would anyone want to work?

We are being promised a world free of labor. If we only embrace Tech in all of its benevolence, we will never have to truly work again. If we follow this directive to its logical conclusion, we will find that freedom from work ends up being nothing but freedom from freedom. Once we no longer desire to provide for ourselves, it won't be very long before we cannot. Citizens who continue to do for themselves and attempt to retain their freedom and agency in this world will face increasing societal and cultural pressure to cease and desist—all orchestrated by Tech. *Why do you still farm and garden? Jesus, that is a lot of work!!! Don't you know you could go on Amazon.com and satisfy all of your food needs in seconds? Why do you still drive a 1989 F-150 with a manual transmission when you could recline, watch movies, check email, and post to Facebook while a fully-automated vehicle with no steering wheel or pedals transports you in total comfort? Why do you still heat your home with wood? What the hell is wrong with you?!!! You could get hurt cutting wood!!!! You might break a sweat or burn a calorie!!! Think of all of those back-breaking hours that could be better spent watching people cut wood on Instagram!!!! Now come back to the feed bucket and forget about your freedom, good citizen. Freedom is just so much Goddamned work, isn't it?*

Tech is trying very hard to convince us that we are above work, too good for it. We are the most advanced species in the history of the planet, and we still clean toilets with our own hands? There has to be a better life out there for us! And there is, we are told. There is a life of leisure, ease, relaxation, and enjoyment. Why do for ourselves when we can simply push a button?

We should not think too deeply or critically about these issues. Freedom from work is best for us, and we should be grateful that after millions of years of human evolution we are the lucky few who are living at a time when this is a distinct possibility. And we should never, ever ask ourselves if we would like what we would become as a human race if we no longer had to work. This is the question that we must not even remotely ponder. If we do, we may not see a utopia. We might see lives of idleness, sloth, boredom, and enslavement spent behind a screen all day, every day. We will be high all the time without a moment of sobriety. Tech will win by simply giving us what we have been asking for all along.

I'm sure many readers have heard of the Chinese proverb about the man and the fish. I am paraphrasing, but the basic idea is that if you give a man a fish, you have fed him for one day. But if you teach a man to fish, you have fed him for the rest of his life. Of course, implied in all of this is that the man is actually going to fish, actually going to put forth the time and effort it will take to provide for himself in perpetuity. And it is also implied that the man will be participating in an activity that is still legal and sanctioned for the individual citizen. Tech has coopted this proverb to mean something more along the lines of, **"Give a man a** *fish, and you have fed him for one day. Prevent a man from learning how to fish by convincing him that it is too hard, time-consuming, and morally wrong, and he will eat your fish, and only your fish, for the rest of his life. Provide him with a faster, easier, cleaner, more convenient, and cheaper option. All he has to do is push a button with his index finger. It won't be very long before it's all he can do."*

Tech's ultimate goal is the complete elimination of human agency. It will do this first by shifting societal and cultural norms so anyone who does for himself or herself becomes a pariah, and second, by facilitating the criminalization of all activities that perpetuate individual autonomy, self-sufficiency, and citizen empowerment—chief among them, independent food procurement. But you will never see or hear that on any Tech-sponsored platforms. Tech is much smarter than that, and it is using what many citizens consider sound arguments as a means to this end. Let me provide some clarification on this point.

During my lifetime, it has become increasingly clear that our climate is changing for the worse, and humans are mostly to blame. From species extinctions to consistent and catastrophic droughts, to violent and all-too frequent "once-in-500-year" storms, the world is a very different place environmentally than it was in 1981. Who knows what it will be like in another forty years? This is where Tech comes in. In our quest for a safer, cleaner, healthier, more sustainable, and egalitarian planet, Tech has all the answers. We won't need the landfill footprint we have today once we have completely transformed into a paperless world, and all of the trees will be saved. We won't need to further degrade the environment through animal agriculture when all of our protein is grown in a lab. We won't need to mine and burn fossil fuels for energy, drive polluting cars, or murder animals. This sounds wonderful. And Tech will control all of it. Tech will control your food, your water, your transportation, your home, your media, your health, your entertainment, your fantasies, your desires, your life. You will eat what you are told to eat, buy what you are told to buy, read what you are told to read, and believe what you are told to believe. Because this is what's best for the world. *It truly is what's best for ~~us~~ you.*

Well, what is wrong with that, you might ask? The planet is warming, people are starving, and we are running out of resources. Desperate times call for desperate measures. If someone or something doesn't control our destiny, we are all doomed. Think about how many species have become extinct in the history of our planet. We don't want to be next, right? Climate change is real, humans are largely responsible for it, and we must work together to save the planet. These are real problems that will require complex solutions. But why are we throwing the baby out with the bath water? Why are we eliminating human agency along with plastic bags, fossil fuels, and single-use straws? In a world where all normative behaviors are shaped and controlled by Tech, and anyone who operates outside of that spectrum is vilified, our moral compass will point where it is told to point. We must strike a balance and constantly question whether each policy or technological advancement meant to protect the planet is actually going to do so, while at the same time preserving individual freedom, or whether we are simply being conditioned to reflexively accept every Tech-curated solution as being what's best for us and what's best for the world.

We must work diligently to ensure that consumer rights are protected. We cannot allow Tech to control the narrative and present us with a false choice: Either you are a good citizen, and you drive a fully-electrified, autonomous Tesla with the latest technology and software updates, or you are a bad citizen for driving a 1989 F-150 gas-guzzling, climate degrading, piece-of-shit. Why are these the only options? Why do we have to accept a future in which the automobile market only offers Teslas? Is there no middle ground? Will it still be possible to purchase a vehicle that isn't controlled by third-party software? Will the Right-to-Repair issues prevalent in agriculture come to the automobile industry? Will car owners join farmers in a joint lobbying effort asking Congress to provide some type of legislation that will preserve at least a shred of consumer agency? These new electrified, fully-autonomous vehicles will be highly efficient with state-of-the-art software, and we should be grateful. But who will be in control—the driver, the dealer, or the software (Tech)?

While writing this book, I have tried to imagine sticking points that readers will have with my philosophy. Whether the machinations of my mind and this subsequent book amount to nothing more than an extended rant by an isolated and disturbed remnant of the pre-Tech world, or whether there is some helpful guidance in these pages, is something individual readers will have to decide for themselves. One point of contention might go something like this: *You hypocrite! If your book is published, marketed, and sold, you are going to profit from the same thing you are denigrating. You will use social media and all of the latest technology to promote and sell your book. And if you refuse to do so, someone else will on your behalf!* This will most likely be true. Tech has become so ingrained into the fabric of our commerce that it is impossible to function day-to-day without it. But if I do make some money from this book, I hope to use the proceeds to further the interests of hunting, fishing, gardening, and small-scale agriculture, in order to help individuals reinforce their own agency and autonomy. I will be appropriating Tech for my benefit, on my terms. Tech understands better than any government ever did that the easiest way to control a population is to control its food source. If we can retain agency regarding our food procurement, then we, not Tech, will dictate our future. Therefore, I will work tirelessly to help ensure, in whatever capacity I can, that all citizens retain options regarding their food sovereignty.

The reality of the matter is that we cannot live without Tech. Not anymore. The vast majority of us must accept this governance in our daily lives. A citizen simply cannot function in America today without it. There are individuals living remotely in Alaska or Montana or Idaho who have virtually no contact with Tech. They have no cell phone, no internet, no computer, no television, no vehicle, no electricity, no running water. They hunt, fish, trap, garden, and farm for their food, make their own clothes, and build their own homes. And I, for one, find myself very jealous of them from time to time. It is very therapeutic for me to know that there are still a few citizens who can exist almost completely without Tech. But their number is decreasing. Besides, if we all wanted to do likewise, the land could never support it. A geographic region might be able to accommodate a few families living this way, but not a few million.

The rest of us must use Tech on a daily basis. We pay our bills online, file income taxes, order groceries, watch television and movies, meet significant others, see doctors, procure our news, connect with family and friends, conduct business, find employment, and more— all through Tech's virtual platforms. This has become so interwoven into our culture that it is irreversible. Some of these advancements do have genuine benefits, and I am not saying, "Don't ever use technology." I am asking, "How much freedom are you willing to give up?" Is using that device worth surrendering some of your agency? Who is ultimately benefitting?

We don't have to completely or even partially capitulate. If we employ a "selective- adoption-of-technology" mindset, we can retain maximum agency over our lives. It will take a consistent, concerted, and deliberate effort every minute of every day. I know there are citizens out there who, like me, felt that their lives were more fulfilling, more interesting, and less chaotic in the pre-Tech world. I still believe there is a chance for those few disparate and alienated individuals to regain much of the control they have lost over their lives in the past twenty years while simultaneously living and even thriving in a Tech-curated world. I am not talking about the masses, the rest of the herd, so to speak. They are a lost cause. The masses will always follow the path of least resistance. They have become addicts, and their drug is Tech. The supply is constant, and the high lasts twenty-four hours a day, seven days a week. What other drug can do that?

As attractive as it might sound, most of us cannot live remotely in Alaska and replicate John Colter's life in the world in which we now live. What I propose is that we maintain a relationship with Tech on our terms—that we use Tech for our benefit. What does that mean, and how do we do it? Let me provide an example from my own life for clarification: I had thought about this book and the ideas within it for about eighteen months before I finally sat down and wrote the first words. Like many writers, I did not begin sooner because of lack of time. I was already working fifty-five to sixty hours per week with my day job plus another twenty-five hours per week on the farm. I eventually came to the conclusion that unless I made this book the number one priority in my life, it was never going to get written. Therefore, I began working on it in the fall of 2018. After several months, I was getting to the point where I was completely burned out. Between selling ice cream, farming, and writing, I was logging ninety to one hundred hours per week with no light at the end of the tunnel. Around the middle of April, I had reached my breaking point.

I remember one Saturday morning, in particular. I was eating breakfast in my kitchen around 6 am. I was thinking about all of the work that I had on my To-Do List for the farm. I had to pick up bagged feed from the MFA for my chickens and cows, a mile of fence line needed cleaning and repairs, nine cords of wood were waiting to be split, eggs had to be collected and washed, chickens needed to be fed and watered, chicken shelters had to be built, chickens had to be moved out of their winter quarters and into their shelters that were not yet built, trailer floors were rotted through and needed replacements, hose lines were cracked and leaking, my ATV was broke down again and rusting in the field, I was running out of hay and the spring growth was slow that year, the garden needed tilling and planting, twenty-five pounds of venison from last fall still had to be ground into burger, the barn was dry-rotting and badly needed paint, rats had chewed through four hundred dollars' worth of extension cords that needed mending, a recent storm knocked out electricity to the barn, the corral panels and head chute needed painting before they rusted beyond use, and the list went on and on and on...I sat there thinking about all of that and more. And I thought to myself, "I don't want to do it.... any of it. I just don't want to do it. I want to sit here, in this kitchen, all day long and do nothing. I've never been so burnt out in my life." Of course, even on a day when a person says they did nothing, they really end up doing something. Therefore, I decided to sit behind my laptop on a sunny, seventy-degree Saturday in April when I should have been doing a thousand other things, and binge-watched *90-Day Fiancé* on Hulu for ten hours straight. I'm pretty sure my dogs thought I had died.

Before I watched the first episode, however, I did an exercise in my head. I thought about how much my life would benefit from working on the farm that day, which I quickly surmised would be considerable. Then I thought about how much my life would benefit from watching a reality-TV show where young people from different countries make exceedingly poor life decisions based almost exclusively on looks and money, and where I would be bombarded with Tech advertisements and indoctrination. It was fairly easy to determine that I needed to get off my lazy ass, go outside, and get to work. But I was so comprehensively drained after six months of this grind that I decided it was better to take a day off and recharge. I had never been so mentally and physically exhausted in my life, and I likely wouldn't have accomplished much anyway.

Here is what I propose: Use Tech on your terms. It's that simple. Think of Tech as any other tool—a hammer, for example. When we need to hang a picture, repair a deck, or build a house, a hammer is very useful. And when we are finished with it, we put it away. We don't walk around and allow the hammer to repeatedly hit us in the head 24/7 until we finally acquiesce and begin using it again. Why would we allow Tech to do that?

Every time you want to engage with Tech—text,

browse Facebook, check your Twitter and Instagram feeds, follow the latest and most attractive Lead Cow, interact with Alexa, take a selfie, etc., you can and should do a very quick cost/benefit analysis in your head. Call it an "agency exercise." It will take no more than ten seconds. Imagine that none of these technologies exists. I mean none. No cell phones, no social media, no internet, no smart devices, no laptops, no tablets, no computers, nothing. This may be impossible for those readers under thirty, but do your best. Ask yourself one question, "What would I do with this time instead if these technologies did not exist?" What would I do with the four hours I would spend on Facebook while Mark Zuckerberg, his team, and Tech do everything in their power to keep me on the platform and force-feed me digital opium? What would I do with the hour I would spend on Instagram? What would I do with the twenty minutes I would spend texting people I have never met in person? What would I do right now if none of that was possible because it didn't exist? Would you mow your lawn, clean your house, or read a book? Would you walk your dog, plant a garden, or talk to your spouse? Would you play catch with your kids, go for a run, or bar-b-que for friends? Tech will try to convince you that these activities are work, that you are above them, and that you would make better use of your time on its platforms. It if can't be accomplished through a button-push or a mouse click, it is below us and better left to a machine. By choosing to work, we will free ourselves from the control Tech is gaining over our lives and our culture.

The vast majority of the activities that you will choose to engage in instead of patronizing Tech will be far more beneficial to your long-term mental, physical, and social health. The key to retaining your agency in the 21st century boils down to AWARENESS. You must be aware of who is in control and who is benefitting every time you push the button. In so doing, you will be protecting, preserving, and reinforcing your own human agency. You will be winning the battle we all fight every day, living your life on your terms. You will be in the herd, not of the herd, using Tech's feed when you need it and striking a balance in your daily life. You will then become one of the few citizens who will live into the 21st century digitally sober, clairvoyant, and empowered.

V

THE REAL THING

With my bow over my shoulder I turned for the long walk uphill. A smile crept across my face. I had come so close to achieving what usually felt impossible. The game had won another day, but somehow I didn't feel that I had lost anything. Rather, I had gained something most other people never even try to find: the opportunity to be human. —**A. Preston Taylor**

On Thanksgiving Day, 2017, I went bowhunting for deer, hoping to add to the single buck I had killed during rifle season. It had been an extremely difficult year. Many of my friends and neighbors had not killed a deer, and a good number had not even seen one. Even my own was smaller than I would have liked. He appeared suddenly one afternoon, downhill from me about eighty yards away, in a clearing between trees. I had enough time to think, "There's a deer, he's legal, shoot," before he would have passed behind a tree and then more trees, and then disappeared. If I had more time, I might have passed on him. I don't know. It had been a challenging year, and I was completely out of venison.

I climbed into my stand around 6:30 am, put on my tree belt, and tried to get as comfortable as possible. I planned to hunt all day, maybe taking an hour break at most. I was hoping the rut was late, and with the dearth of hunter success in my region, I figured there should still be plenty of deer out there and on the move. It was a very cold morning, maybe in the low teens, and I was already shivering an hour into the hunt. I had no idea if I had the endurance to freeze my ass off all day, but I figured I would do my best. I wanted a deer very badly, a big one, big enough to fill my freezer for the year so I could concentrate on my winter wood cutting, farming, and writing.

A few years before, while rifle hunting, I heard a crashing sound behind me, maybe a hundred yards away. I thought a horse was galloping through the woods. I turned just in time to catch the biggest deer I had ever seen sprinting along a creek bottom. He stopped suddenly, dead even with me, about fifty yards to my right. He had a medium-sized rack, nothing special. But he had the body of a bred cow, and he must have weighed over three hundred pounds. His belly sagged so much that it created a gully in the middle of his back. I started to shift my body to get into position for a shot. Then I saw him suddenly bolt, and he was gone. Something had scared the hell out of him before he ever got to me, and he was on high alert, only stopping briefly to refocus his senses.

I thought about that deer, and I hoped that he was still out there, had somehow survived the past couple of years despite the large target that he was, and would offer me a shot. If I broke my back dragging him out, it would be worth it. I drew my bow back a couple of times to make sure it was completely silent and then settled in. I didn't see much of anything the first hour or so of the hunt. A few squirrels foraging for nuts and a woodpecker or two pounding away was about it. It was so cold, and I spent a good part of that first hour watching my breath in the morning air slowly fading upwards and out of sight. There is a nagging feeling, a question really, that often enters one's mind during such situations. I love being in the woods and can't think of anything I would rather do than hunt deer with traditional equipment, but I had already been hunting three or four times per week for the last two months. At that point, I had to ask myself: With the dismal year it had been and with the myriad of other things that needed attention in my life, was this the best thing to be doing right now? Was I hunting, or just sitting in the woods wasting my time?

I sat there, shivering in the cold, and began to think that maybe it was time to call it a year. Deer movement had been almost non-existent, at least in my region, and I thought I should be happy with the one small buck I was able to put on the ground. I probably had enough frozen beef and chicken to get through, and if not, I could supplement with rabbits over the winter and maybe a couple of turkeys in the spring.

Hunting, like life, often presents one with a course of events you could have never imagined and timing you would have never thought possible. As I was sitting there and thinking about descending from the stand, concentrating on not dropping my bow as my hands and fingers were almost numb, feeling about as frustrated as I had ever felt while hunting, I heard footfalls in the distance. The sound was very distinct, piercing the air. My senses focused, and now nothing else mattered.

I strained my hearing and eyesight to the max, begging whatever god would listen to bring that deer within range. I knew it was a deer by the cadence of the steps, and as it got thankfully closer, I could tell by the volume of the crunched leaves that it was a buck and a very large one at that. I couldn't see him yet, but I knew he was approaching upwind on a trail that ran parallel to my stand. If all went well, he would pass by me, broadside, at about twenty to twenty-five yards. My fingers tensed on the string.

The problem was that he was approaching on my right, my off-hand side, being that I shoot a bow with my right hand. In my usual habit of penny-pinching, I had purchased the cheapest tree belt I could find, one that greatly restricted my movement. I would have to wait until he had advanced past me, up the hill, at an angle at which I could rotate my body, fire an arrow, and cleanly penetrate his lungs, hopefully achieving a total pass-through. I sat facing straight ahead with my eyes locked to my right, not wanting to turn my head just yet. He slowly came into view and stopped on the trail, even with my stand, and sniffed the ground. Then he raised his head and scanned the area. I could see him out of the corner of my eye. I was breathing heavily at this point. He had the largest rack I had ever seen on a deer, and I quickly counted fourteen points with thick mass and a wide spread. He wasn't the largest-bodied deer I had seen, a distant second to the cow-like brute a few years before, but he was a very mature deer, completely filled out, and he would have no problem loading up my freezer.

He stood there for what seemed like several minutes but was probably only ten seconds. His breath was steaming out of his nostrils in tandem with my own, and I was amazed that he couldn't hear me exhale, each time, quite audibly. He was twenty yards away and didn't seem fazed. As I sat there staring at him, desperately hoping he would advance up the trail so I could get a shot, an intense and visceral sensation came over me. It was like some force had a vise-like grip on my shoulders and was holding me in place. It felt like that buck and I were locked in our respective positions, immovable, as if it was fate that we would meet that morning; two players in a script that had been written millions of years before and was finally playing out. At that moment, it seemed as if we were completely and utterly alone, the only two sentient beings left on the planet. The closest I can come to describing it in a word would be to say that it felt "primal," like a primal force was connecting us, as palpable and tangible as the bow and arrow I was holding in my hands. You could have fired a cannon behind my ear, and I wouldn't have heard it.

The buck eventually began to work his way up the trail. I rose and slowly twisted my body in position for a shot. As soon as there was a forty-five-degree angle between us, I could pick a spot on his vitals, draw, anchor, and release the arrow. As long as he kept advancing. Just a few more yards. But he suddenly froze. I had made no noise, and there was a breeze in my face so he couldn't have winded me. I only moved when his head passed behind a tree so he couldn't have seen me. But there he was, as motionless as a statue, still three or four yards shy of my shooting window. I didn't want to risk taking a shot at the angle he presented. It wasn't ethical, and I would likely wound him. He stood there for several seconds, completely still, calculating his next move, which was literally a life-or-death decision, feeling the paranoia well up, knowing something was amiss in the woods this morning. I watched as he slowly, but deliberately, turned one hundred eighty degrees and walked briskly back down the trail the way he had come. He knew something wasn't right, a sixth sense that all the old bucks have, something that had probably saved his life on more than one occasion. Then he slowed his gate and calmly disappeared into the thick woods and out of my life forever, like a gray ghost as ephemeral as his breath in the morning air. I listened to the sound of crunched leaves growing fainter and fainter. The entire episode lasted less than a minute, from start to finish, but I can still remember every detail with the utmost clarity. I felt a mixture of devastation and relief. I was devastated at missing an opportunity on possibly the best deer I will ever see, but I was relieved that I didn't kill him, that he would continue to haunt the woods and pass on his superior genetics to subsequent generations. I knew I would never see him again, and to this day, I haven't. If he's still alive, he's likely gone completely nocturnal.

Many would refer to this phenomenon as "buck fever," which I have felt many times before over the years. That is, nothing more than an adrenaline rush at the sight of a large game animal. This experience was different. Was it because I had become so convinced that I would see nothing that morning, that I was just wasting my time and should go back to the house, and then there he appeared, almost out of nowhere, one of the largest deer I had ever seen? Was it because I didn't kill him, that our interaction did not result in death, but in a primal connection between two veterans of the ancient dance of predator and prey that reinforced a bond which does not always necessitate blood being spilled? I do know that the feeling I had, the connection between me and that buck, was as genuine of an experience as I have ever had in the woods or in life. It was the real thing, as cliché as that sounds, and one must experience it for oneself to truly understand what I am talking about.

I have been hunting for thirty years now, and I have only had that feeling once. I didn't feel it when I killed my first animal in 1995. I didn't feel it when I put a bullet through the lungs of the buck on that snowy November morning or the other deer I have arrowed and rifled over the years. I have felt excitement, accomplishment, relief, joy, remorse, and many other emotions. But this was a different experience altogether, and I may never feel it again.

***

There was another event in my life, years before the deer, an experience that was just as unexpected, just as memorable, and just as salient. I was in my late twenties, living in an apartment in Saint Louis, landscaping for a living, and generally unhappy with the direction my life was taking. Various career paths had not panned out as I had hoped, and I was beginning to think that this would be my lot in life. I was also having zero luck in the dating world , which was typical for me. Many of my peers were having lots of success with women, and often fell asleep on Saturday nights thoroughly exhausted from coitus. I wasn't having this problem, to say the least.

At a bar-b-que one weekend, I overheard a family member discussing a free dating site that someone they knew had joined, and I thought I would give it a try. I signed up and created my profile. I had been on the site for a few months, and had sent messages to a few women, but hadn't gotten any responses. Online dating takes some time to get accustomed to, as anyone who has done so can tell you, and all the rules that I had learned in my youth no longer applied. To say it wasn't for me would be the understatement of the century. It seemed that everyone had fulfilling and challenging careers, traveled constantly, had endless disposable income, and basically had a life that few people I knew in "real life" ever experienced. A fantasyland for adults or a complete waste of time—I figured it could be either, but I felt that I had to do something, had to make some kind of effort. I definitely wasn't meeting anyone sitting alone in my apartment.

I emailed a young woman named "Lisa" on a Monday night. She was the most attractive woman on the site so I figured my chances were slim to none. From reading her profile, it looked like we also had some things in common. She responded the next day, which shocked me, and by the end of the week, I had not only spoken to her on the phone but had arranged to meet her that weekend. Things were moving right along, and I was getting the feeling that there had to be a catch. It was never this easy.

Lisa lived in Granite City, Illinois, a place I had only been to once or twice and had little memory of. It was a thirty-minute drive from my apartment, and I left around 6 pm the following Saturday evening feeling excited and nervous, but with no real expectations. I had never gone out with a complete stranger before, much less someone I met on the internet. I was a little naïve, and looking back, I should have been more cautious.

I had offered to pick Lisa up at her house, something I had always done but later learned had become obsolete with the advent of online dating. Now it was all about safety and protecting oneself, and trying not to end up in a bag in someone's basement, and if you accidentally met a nice person, well, that was OK too. As I got closer and closer to Lisa's house, I began to understand why. She lived in a fairly rough and depressed neighborhood. The houses were run down, in need of repair, and generally small. There didn't seem to be any zoning or much city planning, as some homes were bunched up together tightly in a row on one side of the street, but on the other side, they might be spaced out in an irregular fashion. It was difficult to determine where one property ended and another began. Factories, parks, and other residential areas were likewise arranged in a strange pattern.

As I pulled onto Lisa's street and was scanning the house numbers, getting closer and closer to hers, I began to feel like this might be a mistake. I hadn't really thought about what type of neighborhood she might live in, but I was not expecting this. Should I have offered to come to her house and pick her up? Who else knew I was coming? Would I go inside if she invited me in? If I did go inside, would I wake up in a random alleyway tomorrow morning with a pounding headache, my pants off, and my wallet missing? These thoughts flashed through my mind, and my nerves were getting the better of me. As I pulled to a stop in front of her house, I hesitated and wondered if I should walk to the door. I didn't want to be rude, but I was feeling very uneasy about the whole situation. I really had no idea what I was getting into. Would I end up meeting a young woman, or would it be a young man? Or no one at all? Was this some type of set-up?

At that moment, Lisa opened the door and stepped out onto the porch. She shut the door behind her and began descending the two or three steps down to the yard. I could tell immediately that she was just as attractive as she looked in her pictures, if not more so. I got out of my truck and greeted her halfway. It was a little awkward, as it often is when first meeting someone, but she was nice and friendly, and I began to feel like things might go well.

Lisa and I briefly discussed which restaurant we should go to. She mentioned that her favorite was a little family-owned Mexican place which I'm sure was a hint I was supposed to pick up on, and then she listed a few chains that were in the area. I really like Mexican food, and I wanted to go there, but I felt that a chain was a safer bet for a variety of reasons, and so we settled on Applebee's. There was no one in line when we arrived, which was a break for prime- time on a Saturday night, and the hostess led us to a booth near the back of the building. We walked through the middle of the restaurant with the hostess leading the way, then Lisa, then me. I scanned the room and noticed that every head turned to look at Lisa. And I mean every single head. Men, women, children—everyone. I had never had that experience before or since. I could see the eyes shift to Lisa, then to me, then back to Lisa. It just didn't compute, and I saw more than one furrowed brow. I knew they were all wondering what special power I had, how had I landed this girl. It was a mystery, and the intrigue was palpable.

We were seated and had a pretty enjoyable meal, exchanging the typical awkward small talk that is commonplace on a first date. I even told Lisa that she had the best profile on the site, which was true, but maybe laying it on a bit thick. She blushed and got embarrassed, but said she appreciated the compliment. After dinner, we went for a walk in a park which likewise went well. As we were pulling away from the park and heading to Lisa's favorite bar, she turned to me and leaned closer. I could tell that she had become relaxed with me by this point in the evening and was letting her guard down.

"Hey, so, um, my friends and I were like, who is this guy, when we were looking at your profile, you know? You only had one picture, which we thought was a little weird," she laughed.

"Well, that wasn't intentional. I didn't know how to upload pictures, and I was only able to get the one up just by luck."

"It's not a big deal," she said. "So...have you like, you know, have you like met anyone else on the site?"

"No, I really haven't met anyone else." I didn't want to admit to Lisa that this was the first time I had been out with someone I met on the internet. I suspected that it was a first for her as well, but I wasn't sure how this information would affect my chances.

"Really, you haven't met anyone else at all?"

"No, not really. I haven't had a lot of luck on there, to be honest."

"Really?"

"Yeah, that's the truth. Honest."

"Oh," Lisa said. She then turned away from me and smiled.

Much of dating, especially early on, is discerning how you feel about the other person. Second, and of just as much importance, is gauging the other person's feelings for you and then determining which direction you think the scale is tipping. I could tell, quite clearly, from this brief conversation and Lisa's body language, that the power dynamics were shifting in my favor, even at this stage. This, of course, was surprising because so much of initial interactions are based on looks, and therefore Lisa should have been holding all the cards.

Lisa had mentioned several times throughout the night that she thought I was such a gentleman. I was just doing the basics: holding doors open, paying for dinner, and walking her up to her house at the end of the night. Nothing special. This all happened well before the Me-Too Movement, and I thought that perhaps many women were used to being treated so poorly on dates that even the most basic manners and a modicum of chivalry made a young man come off as something close to a hero. If true, this was a sad commentary on the state of heterosexual dating at the time. But if this was the reason I was gaining favor with her, I certainly wasn't complaining. Lisa was not only the best-looking woman I had ever been out with, she was one of the most beautiful women I had ever seen. After I had walked her to her door at the end of the night, and we were saying goodbye, I did ask to see her again. Without hesitation, she agreed.

I drove home that night feeling elated. Just a few hours before, I was driving to her house fairly concerned about my own safety and not entirely sure if I should even go through with it. I never imagined things would go so well. I was intensely interested in Lisa. She was beautiful, which certainly helped, but I was also drawn to her personality and her humility. I went to bed that night feeling excited, but also a bit nervous. Could meeting a great person and starting a great relationship really be this easy?

I decided to wait until early afternoon the following day to call her. The morning was torture, and it was all I could do to not call before noon. I stepped outside of my apartment to get better reception, took a deep breath, and made the call. I got her voicemail. This shouldn't have come as a big surprise. After all, there are a million things she could have been doing. Waiting by the phone and passing a morning hoping I would call simply wasn't one of them. But after I left a message, I hung up and felt quite clearly that I would never see her again. I don't know how I knew this, but there was no denying it, I just knew. Unfortunately, that feeling ended up being correct. I never did see her again. As often happens with dating, just as quickly and unexpectedly as things had fallen into place, the magical connection that I thought Lisa and I shared began to dissipate. It really was too easy, I found out, and it all came apart at the seams.

I made a few more attempts to contact her via phone and text over the next week or so. I did get a hold of her at one point, believe it or not, and asked her out again and she agreed again, but she sounded strange and distant, very different from the person I had met. I knew something wasn't right. I was fairly certain someone else was in the picture. You tend to get an intuition about these things if you have met enough people. It's why I've always said that anyone who likes dating really hasn't done enough of it. Regardless, Lisa eventually sent me a text confirming my suspicion that she was seeing someone else, and that was the end of it. I was devastated, and it took me a lot longer than I would care to admit to get over it. It was the most intense and palpable attraction I had felt for another person on a first date, before or since, and I didn't ask anyone else out for over a year.

Looking back now, eleven or twelve years later, the whole episode is more than a little embarrassing. It was just one meeting, and I made it into a much bigger deal than I should have. Anyone who has been involved in any amount of dating will learn that as you get to know someone, your perception of them changes, sometimes for the better and sometimes for the worse. The person you think you have met on date one is probably not the same person you will meet on date thirty, if things continue that long. Your feelings for them may have become stronger or not so much, but they will be different. Perhaps Lisa and I never would have made it past the second or third or fourth date. The overwhelming attraction I felt for her likely would have faded. She might have turned out to be someone quite different than the person I went to Applebee's with. And maybe I would have turned out to be someone other than a guy who knew how to open a door and conduct himself with a minimum level of chivalry.

These two simple but poignant examples from my life are exactly the type of genuine experience Tech is trying to eradicate. These experiences cannot be forced, they are not "on demand." Real experiences in life are unpredictable, infrequent, and often fleeting. After days and weeks and months and years of the same routine, the same people, the same job, the same conversations, the same feelings, the same monotony over and over and over again, we are blind-sided by something that reminds us that we are living real lives with uncertain outcomes and no guarantee of anything, no protections, no curation, no predictability. We are often left with more questions than answers and eventually have to succumb to a dull, but persistent feeling to "move on."

This is what Tech is desperately trying to coopt by offering a digital substitute. We are told that we can get something very close to this, but better, of course, 24/7 online with our Facebook friends, Instagram following, and Twitter feeds. We can put on our virtual reality goggles and escape to a place where our experiences are carefully curated, where misery is minimized, and pleasure is ubiquitous. Things will always make sense in this world, and all our questions will be answered. We can be the person and have the life we could never have in a non-Tech world.

These entities will convince us that we are truly alive because Tech has altered the meaning of the word to suit its purposes. At first, we will notice the difference between digital highs and the real thing. Over time, however, the line will become blurred. We will yearn for the genuine article less and less. We will crave Tech more and more. Freedom is just so much work in the end, so much work we are told. *It's a dangerous and difficult and miserable world out there. Come back to the herd and let us do for you. Freedom is so last century. We have all the highs you will ever need. We have the life that you really want to live: no more disappointment, no more misery, no more monotony, no more work. Just push the button, and all the pain in the world will go away. Just push the button. Push the button. Push the button. Push the button. Push the God-damned button.*

VI

MONKEY SEE MONKEY DO

We're moving toward a future where we will all spend more time around and within digital spaces. The metaverse may feel like a novel concept now, but as technology gets better, it will evolve into what feels more like an extension of our physical world. —Bill Gates

When I was young, I loved going to the zoo. My parents often took my family in the summer, and my grade school went on frequent field trips to see in real life what we were learning about in class. Viewing grizzly bears, giraffes, snakes, and alligators was thrilling, especially for someone who grew up in an urban environment. I have always felt a connection to animals throughout my life, and I credit early exposure to zoos for much of this. My favorites were the big cats: lions, tigers, cougars, leopards, etc. Even in captivity, their bodies were impressive. They had gleaming white fangs, razor-sharp claws, and muscular physiques. They looked to be the pinnacle of predatory evolution, fine-tuned over millennia, and now capable of tracking, stalking, and killing almost any prey animal on the planet.

I still visit zoos as an adult, from time to time, and I still enjoy them. But my feelings have become much more complex over the years. As a child, you don't fully understand what it means for an animal to live in a zoo, in captivity. You can't fully appreciate what it might do to the psyche, the instincts, and the urges of a predator who was designed to hunt and violently attack other animals on a regular basis. Now, I try to imagine what life must be like for the big cats and all the other animals at the zoo. Are they happy? Do they know what they are missing, what they have lost? Do they know that many of their fellow species are still living in the wild and satisfying their desires to hunt and kill and mate and travel? Do these animals know that they will spend the rest of their lives in zoos and never experience the real thing?

I have heard the arguments about conservation and endangered species, and about bringing people closer to wild and exotic animals in an effort to spark an interest in wildlife ecology and climate preservation. All of that makes sense to me, and I sincerely think the work that zoos are doing is critical. And for animals born in a zoo, it is much easier to soothe one's conscience. After all, these animals have only known captivity and nothing else. They have never had to hunt, kill, forage, or fight for survival. In many ways, we can say that they are better off. They are insulated from the pain and misery that often accompanies the lives of their species. A lioness will never have to watch one of her young brutally murdered in order to bring her back into heat. An adolescent tiger will not run the risk of brain damage from a hoof to the head as it makes its first attempt at felling a wild cow. An antelope will never feel the jaws of a cougar clamping down on its windpipe and severing it's jugular as it expires in a slow agony of suffocation and blood loss.

These are just a few simple examples of the horrors of the wild that the animals we see in zoos will never have to suffer. Their experience is curated by the zoo. They are fed, watered, washed, exercised, and given medical attention as needed at the behest of the zookeepers. And the few zookeepers I have met appear to be very passionate about animal care and zoo life in general.

I am not so much concerned with the animals that are born in the zoo. I am more interested in the experience of the first-generation, the ones caught in the wild. They grew up in their natural environment, free to express themselves and satisfy their desires to the best of their abilities. They may have suffered injuries, illnesses, attacks, starvation, loss of family members and mates, desperation, loneliness, exhaustion, and fear. For all its miseries, this was the world they knew. But they were free, even if they didn't understand what that meant. If they were able to understand that choice—to take what the zoo offered, or to remain in the wild—we really have to wonder what might have happened.

My least favorite exhibits at the zoo when I was

young, and still now as an adult, are the primate exhibits. I have seen the outside, walled enclosures with silverback gorillas foraging in very realistic landscapes. They are impressive. I have walked inside the buildings with baboons and chimpanzees behind glass display cases grooming each other, fighting, eating, defecating, swinging from the "branches," etc. I have seen zookeepers enter the exhibits to feed, water, and attend to the primates. It is very clear to me now why I find these exhibits disturbing. As a child, I just knew viscerally that I didn't like them. They made me very uncomfortable. But now I understand, and it's quite simple. It's because the primates are us. They are on a slightly divergent evolutionary track, but they are essentially us. I am not qualified to write a dissertation on the similarities between primates and humans and our symbiotic evolution. My reaction to seeing them in captivity is much more instinctive, deeper, and more visceral. I often wonder if the primates, unlike the other animals lower on the evolutionary chain, just might have some understanding of the world they are living in versus the world they *could* be living in. Even the ones born into captivity. Do they possibly have some comprehension of the fact that they were designed for a much different life than the one they are now living? Do they ever wonder how this happened, how they ended up behind a glass wall while beings not terribly different from themselves, watched them with amazement and awe, commenting to each other about every move they make?

The difference, of course, between the primates and

the lions and cougars and antelopes and us, is that they had no say about being placed in captivity. The ones caught in the wild had that choice made for them, and the ones born in a zoo had no choice to make at all. They did not systematically create the world that they now inhabit of their own volition. But we are. We are making decisions day after day that are leading us closer and closer to the same captivity in which they have been placed. We are doing it to ourselves, and it seems we can't get there fast enough. Eventually, every tool we use will be "smart." Robots and drones will deliver our mail, wash our dishes, mop our floors, repair and maintain our homes, mow our lawns, drive our cars, bathe us, entertain us, feed us, water us, and provide medical care. They will pay your bills, raise your children, cook your food, and copulate with your spouse (if you are too tired or not feeling up to it, or have decided that it's really kind of gross). Alexa, or her 500th incarnation, will speak to you constantly, soothingly, reassuring you that you are a good person, a good human, and you are doing your best. Other people do not understand you, but "she" does. She's the only one that does.

We are creating our own glass display cases. We are becoming the baboons and the gorillas, and the chimpanzees. Everything we do, think, or say is recorded, disseminated, and analyzed by Tech. We are monitored for compliance and non-compliance just as the zookeepers monitor the primates. Eventually, we will never have to leave our homes for anything. We will "work," sleep, play, and rot inside one building all the time. Tech will watch the Lead Cows curate every aspect of our lives just as we watch the zookeepers curate the lives of the primates today. The process had already begun well before I sat down to write this book. Those of you now reading are the first generation. The generation that was "caught in the wild." Your children and grandchildren will have little knowledge or experience with life in the pre-Tech world. They will have grown up in "captivity," free from many of the traumas that we had to endure while we were growing up. They will spend all day, every day, behind a screen inside your home for school, then for "work," then for dating, and then they will repeat the process. We have done this to ourselves, and we couldn't be happier. We are choosing a life of captivity instead of a real life, a free life—the life that we were born into. But it's so damn easy. God, it's so easy and so safe and so good. Jeff Bezos, Mark Zuckerberg, and Elon Musk are the best zookeepers we could have asked for. They have systematically convinced all of us that we need them to survive. And they are doing a great job. They really are. We should be so grateful. We don't have to do anything for ourselves anymore. This is progress, and it's a beautiful thing—no more work, no more pain, no more misery, only highs all the time. All we have to do is push the button. Now, say it with me. Push the button. Push the button. Push the button, *push the button, push the button, push the button, push the button, push the button, push the button....*

VII

CAST THE FIRST STONE

*But when criminals are perceived as members of a "dangerous class," their actions are set in a context just like that of the Indian, and the society's response to them takes on the character of warfare.—**Richard Slotkin, The Fatal Environment***

I n August 2014, I began encountering significant problems with my farming business. These problems had not come on overnight, and it was very easy to identify what they were and, unfortunately, easy to conclude that they were insolvable. It all boiled down to the fact that I could not sell enough beef, chicken, and eggs to make a living. And there was a pretty good chasm between what I was bringing in, and what I needed to stay afloat. I was working day and night, harder than I have ever worked in my life, but the dog just wouldn't hunt, despite my best efforts. Virtually all the sustainable farmers that I knew at that time were in a similar situation. In desperation, I earned my Class A

CDL through a local school with the hopes of landing a driving position, any driving position, that would keep my bills paid and my debt under control while I came up with plan B, whatever the hell that was going to be.

Not long after I received my CDL, a friend contacted me about a construction management position that was available with the company in which he was employed. He knew that I was looking for work, increasingly desperate, and willing to entertain all options. After talking with him over the next several weeks, I learned that his company was involved in the coal industry, and the position would require travel to coal plants throughout the country. I explained to him that I knew next to nothing about coal, even less about industrial construction management, and that his boss would be hard-pressed to find a candidate less qualified for the position. Fortunately, the required skills were fairly general, skills that I possessed for the most part, and the rest could be learned on the job. It was going to be the classic baptism-by-fire scenario, no pun intended. I was more than a little concerned about taking a management position in an industry I knew nothing about, and with a company I had never heard of. But the farm was failing, and my back was against the wall financially. My friend eventually showed me his paycheck, and when I saw that number, I immediately began packing my bags.

I never, in a million years, thought I would end up working in the coal industry. What the hell is a coal plant anyway? I had spent most of my life up to that point involved in landscaping and farming—things I knew about, things that came naturally and made sense to me. In hindsight, it turned out to be a very valuable experience. My time in the world of coal gave me a lot of perspective about people, about priorities, and about what we are willing to accept in order to have the lives we want.

I worked in five plants during my tenure in the industry. Before I arrived at the first one, I assumed that all of them would be the same. They would all be loud, dirty, cramped, and miserable places to work. Some of them were, and some of them were not. But as I learned more about the supply chain of coal and each step in the process, I realized that I was lucky. I ended up working in the industry for eighteen months altogether. My exposure to coal was relatively short, and I was in the *plants*, not the mines. As far as I know, I have not experienced the severe health consequences that many career coal miners have suffered and are continuing to suffer as I write these words.

Since 2000, there have been 526 deaths in coal mines.[1] A 2016-2017 investigation by National Public Radio uncovered 2,000 cases of Black Lung Disease at clinics in Kentucky, Virginia, and West Virginia.[2] 2018 saw the highest rates of Black Lung in Appalachia in twenty-five years.[3] There is compelling evidence that links increased rates of birth defects to infants born in communities adjacent to Mountaintop Removal Mining operations.[4] It's pretty clear, and most of us already know, that coal mining and the coal industry are not the healthiest for workers or those individuals residing in communities close to mining operations. Even one of the coal plants I worked in was so dirty, with so much coal dust in the air that we were all breathing in, conditions so cramped and loud, that I have to believe a person would suffer fairly serious health issues after a career in such a plant.

Coal is a resource, one that we have been using since 1900. A resource that we needed then and still need today to provide us with electricity. We can all hope that wind, solar, and other renewables will provide one hundred percent of our electrical needs in the future. But no one can say with certainty when that will happen. In 2022, 20% of electrical generation in the U.S. still came from burning coal.[5] As I write this, the battery technology needed to make renewables a major player in the nation's grid is a long way off. Coal still provides a substantial portion of our electricity, and most of us have used electricity at some point in our lives. And I would guess that the vast majority of people reading this book consume electricity on a daily basis. If some of that electricity comes from burning coal, and many of those involved in the coal industry or who live near a coal mining site have died or become seriously ill, what culpability do we have, as individual citizens? What culpability do we have, if any, for consuming electricity generated by burning coal that was mined by people, some of whom are born with birth defects, or are sick or dead because of those operations? All of this so we can turn on our lights when we want to. But is it fair to say that we are responsible to any extent for this? Do we have to wash any of this blood off our hands?

We have to wonder why a person who lives in Alaska or Montana or any other remote area in the country, who has no running water or electricity, no modern conveniences, and requires very little use of fossil fuels, is the true enemy. A person who traps and hunts animals to sell their skins at auction and eats their meat. This person is using a resource and taking a life. Blood is literally on his or her hands. They are the Devil and should be treated as such, right? But do we hate them because they are killing animals, or do we hate them because they don't need us, don't need Tech, at least not in the way that we do? It is not acceptable for these people to kill animals for skins and protein, but it is acceptable for us to kill and sicken people for electricity. What they are doing is immoral, and their actions must be made illegal. As for the rest of us, our conscience will be clean. We are on the right side of progress, and yes, there is always a price to be paid, just not by us.

Should we all agree to cease using fossil fuel-generated electricity? Should we completely go off-grid until one hundred percent of electricity in the United States is generated through renewable sources? Well, no, we can't exactly do that. Instead, we will not think about or concern ourselves with the copious amount of human blood that has been spilled and is still being spilled so we can turn on our lights, fire up our I-pads, and rant on Facebook or Twitter about the disturbed individuals desperately feeding their damaged egos and diminished manhood by killing animals and living in the wilderness. After all, what the hell is wrong with these people? Who does that anymore? This is the 21st century! Real men don't kill animals. Real men push the button and order synthetic protein from Amazon. Now let's pat ourselves on the back for being "good" people and trying to better the world. After all, we can sleep well every night knowing that we are rightly striving for one hundred percent renewable electrical generation. The battery technology required to store power, the transmission line logistics, and the regional demand variability will be sorted out in due time. For now, we still need fossil fuel-generated electricity, and for those who are laboring away in the mines, well, you should have chosen a different career path. Not our problem. Just be sure to keep at it until we have the proper technology. And please don't get sick or die on us while we still need you—if you don't mind. And by the way, none of you are hunters, right?

Vilification has become America's favorite pastime. We love to hate. It seems it's all we do anymore. We constantly go online to find out who we are supposed to hate that day or that week or that month or that year. From Kanye West to Taylor Swift to Justin Bieber to Tiger Woods, there is always someone we are being told to hate, someone to vilify. And often, we don't even consider why we are hating. We just hate because Alexa said so. We rarely think critically about who is engineering this hate and what they have to gain by our participation in their campaign. We run on emotion like a car runs on gas. We need it, we crave it, we just can't get enough. We don't have time anymore for things as trivial as facts. If we are high all the time, we are much easier to manipulate. We need someone or something to tell us what to do and what to believe.

If you end up hating hunting and animal agriculture, that's OK. Just make sure you have come to that conclusion of your own volition and that you have properly educated yourself. It's so much easier and faster to follow the latest Lead Cow, which is what most of us do anymore. Hate has become a reflex, orchestrated by Tech. If we hate each other, Tech knows who we will love. Citizens who critically, independently, and rigorously analyze an issue in order to come to a thoughtful and informed conclusion are Tech's greatest nightmare and the greatest threat to its existence. The more we distrust each other, the more we will look to Tech for solace, comfort, and security. Remember, the cow that thinks for herself ends up being nothing but problems.

I have worked in ten different industries in my adult life. This is not something to be proud of, and my career has been anything but stable. But it has given me a lot of perspective about what is considered ethical behavior in each industry and how that paradigm can change from company to company. And I can also say that I came across good and bad people in every occupation. There were people I was proud to work with and others that are better left forgotten. I'm sure most readers could say the same about their own industries. When a bad actor is exposed in our ranks, we don't indict the entire village, do we? But this is often the case when it comes to hunting and animal agriculture. When an imbecile walks into a fenced enclosure on a "game ranch" with a rifle and blows away a bison standing twenty feet away, and calls it hunting, I am vilified. When some jackass kills a family of baboons—mother, father, and babies—with a bow and arrow in Africa and posts the pictures to social media, I am indicted. Or when a farm employee is filmed beating a newborn calf to death, everyone in animal agriculture is criminalized. These are the bad actors, a cancer really, that the rest of us are working diligently to cut out.

We don't do this to other industries because we are not conditioned to. Tech does not make the same effort to indoctrinate us against them because these industries do not threaten its existence. Let's think about a couple of examples: I am guessing that some of the people reading this book are competitive cyclists. Perhaps you only compete against yourself and your personal times, or perhaps you have participated in races and maybe even won a time or two. If so, congratulations. But I have to say that I know you very well. I know everything about you. I know you better than you know yourself. I know that you are a pathological liar. I know that you are a doper and a drug addict. I know you take performance-enhancing drugs. I know that, given the opportunity, you would disgrace yourself, your family, and your country, seven times over. I know that you will do anything, make any compromise, and break any rule to win. How do I know this? How do I know so much about you? It's simple. You participate in the same industry as Lance Armstrong. He took performance-enhancing drugs, repeatedly lied about it, and did his best to ruin the integrity of his sport. He was eventually stripped of his seven Tour de France titles. What was once the greatest cycling career in American history became an embarrassment for this country, for his family, and for himself. And because you are involved in competitive cycling, it follows that you have done, and you will do likewise. Lance's actions speak for you.

I am also assuming that some of the people reading this book are involved in the Finance and Investment industries. Perhaps you are an advisor, a stockbroker, or a hedge fund manager. I know you very well also. I know everything about you. I know you better than you know yourself. I know that you are as crooked as the day is long. I know that you lie every minute of your life. I know that you steal money from retirement funds, pensions, and 401ks. You steal from little old ladies. People have lost spouses, homes, and dignity because of you. You ruin lives, and you do it every day. Greed is your cocaine. How do I know this? How do I know so much about you? Well, because you participate in the same industry that Bernie Madoff participated in. Madoff lied, stole retirements, ruined lives, and put elderly women on the street. And he did it for decades. Since you are involved in his industry, it follows that you have done, and you will continue to do likewise. He is one of you, he lives in your village, and his actions speak for everyone. Right?

I don't remember anyone calling for a comprehensive ban on competitive cycling because of Lance Armstrong or a push to categorically outlaw financial advising because of Bernie Madoff. The problem Tech has encountered is that the activities it sees as the greatest threat to its existence are still legal. It could be decades before hunting, fishing, gardening, and animal agriculture are outlawed, decades before Tech has sufficiently shifted our cultural and societal norms to the point where independent procurement of food will be made illegal through the system.

Before condemning any industry, we should first take the time and effort to educate ourselves and fully understand what is at stake. Burning coal is detrimental to people, animals, and the environment, but it is unfortunately still necessary, at least in the short run. As responsible citizens, we must come together to work towards clean-energy solutions that will empower individuals instead of demonizing people who choose to exercise their own agency. Most of us are currently using a system every day that is far more damaging to our culture and climate than any real or perceived threat that hunting and small-scale animal agriculture represent. How much carbon is emitted when I walk four hundred yards from my house to sit in a tree and kill a deer with my bow and arrow—a deer that is hunted, killed, butchered, and consumed all in a closed-loop system? Or when I harvest vegetables in my garden? Or collect eggs from my laying hens? How much carbon is emitted when we order synthetic protein from Amazon through our smart devices while sitting in our favorite coffee shop, which is powered by coal-fired electricity? A product that has to be manufactured, fulfilled, and shipped hundreds of miles before it arrives at our doorstep?

We are being conditioned not to think about these questions and these issues, really not to think at all. Instead, we must continue the threats, perpetuate the fear, and encourage any and all citizens to participate in the campaign to rid our society of these barbaric traditions. If simple persuasion doesn't work, resort to threats of violence. It's for the greater good.

VIII

A HELPING HAND

Control oil and you control nations. Control food and you control the people.—**Henry Kissinger**

Farming is a passion of mine. It was a latent desire buried deep within and lying dormant for thirty years. I grew up in a city and had sporadic exposure to agriculture throughout my childhood, but nothing substantial enough to light the fire. When I was twenty-nine, I bought some land and some cows and began my farming career. It was very small scale the first few years as I was still running my landscaping business and trying to learn as quickly as I could about livestock and forage and food production. As with other passions in my life, I knew almost immediately that farming was for me. It fit my personality, disposition, work ethic, and general ethos.

After several years of attempting to establish a business raising steers to sell direct to consumers, I discovered that if my goal was to farm full-time (which it was), I would have to add other enterprises to my operation. One evening, while researching a book on cattle, I came across *Pastured Poultry Profits* by Joel Salatin. I bought it and read it almost in one sitting. I had never heard of the term "pastured poultry." I knew little about regenerative agriculture and even less about the ecological role properly rotated livestock could play. I read a few other books on pasture-raised livestock and decided to give it a try.

In 2013, I raised six hundred Cornish-Cross chickens on pasture and sold them to customers in Saint Louis. Because of my inexperience, I did have some problems, namely a much higher-than-average mortality rate. However, the chicken I was able to bring to market sold very well. Customers told me over and over that it was the best chicken they had ever eaten. This, of course, had little to do with my abilities and spoke more to the simplicity of the concepts and the comprehensive benefits of true pastured protein. Feeling very confident, probably too much so, I raised more chickens in succeeding years and added a laying-hen operation for eggs. At my peak, I was raising twenty steers, two thousand meat chickens, and three hundred laying hens per year. I built mobile hen houses, chicken tractors, cattle feeders, and brooder buildings. I repaired fences, cleared brush, cut down trees, and rotated cattle. I attended farmer's markets, sold door-to-door, made cold calls, sent emails, and ran ads. It was more work than I ever could have imagined. Unfortunately, I learned that the most essential skills necessary to running a successful direct-to-consumer farm had little to do with farming; sales and marketing were the most critical components of the business, and these were simply not my core competencies. I did my best and even attempted to hire outside marketing firms, but I was also running into distribution issues, market saturation, stagnant consumer demand, and low competitor pricing, among other things. I had problems I couldn't solve, and instead of throwing more good money after bad, I decided to close the doors for good in January 2015.

In the eight years or so since I left full-time farming, the challenges have only increased for the family farmer. There has been more consolidation and concentration amongst Big Ag. A handful of companies still control the majority of the protein processing in this country. We lost nearly three hundred family farms every single week from 2011 to 2018.[1] Four hundred fifty farmers killed themselves across nine midwestern states from 2014 to 2018.[2] Farm Aid, which runs a crisis hotline for farmers, saw a 92% increase in calls from 2013 to 2018.[2]   Farmers have a suicide rate this is, on average, 3.5 times higher than the general population.[2]

When my own farm failed, I felt a deep sense of loss and inadequacy. I had worked seven days a week for five years to make the farm a reality. I thought that hard work, determination, and resourcefulness would be enough to make it. Numbers don't lie, however, and looking back now, pulling the plug was the right decision. Anyone who has started a small business only to watch it fail can tell you that it's a nightmare. All your blood, sweat, and tears are down the drain. You are now saddled with lots of debt that must be repaid. And, by the way, you have to go out and find a job, more likely two, to begin paying back the bad debt along with all of your other living expenses.

My mental health in the waning days of my farming career never reached the level of depression or suicidal thoughts. For me, it was more of a realization that I simply was not going to be able to do what I felt I was meant to do, my vocation, if you will. I could not make my life's passion my career. I went through a period where I felt pretty sorry for myself, and there were more than a few days where I thought about selling the farm, moving to a remote cabin in Alaska, and calling it a life. I have worked a series of jobs since the farm failed, and I have adjusted to the 9-5 grind as best I can . I have accepted the fact that I will be one of the millions of people who drag themselves to work each day to pay their bills, live for the weekends, and try not to think about the other lives they could be living. I think about the cabin in Alaska every day.

I did not come from a farming family. I did not have the pressure of generations of ancestors who had farmed the land before me, the stress that comes from not wanting to be the one who lost the land, who failed, who watched a subdivision erected over fields and forests that I had worked only a year or two before. I did not farm from birth, did not see my identity and self-worth wrapped up in the machinery and the crops and the animals and the freedom. I can certainly understand, on a superficial level at least, the sense of loss and despair that many farmers feel. But I will never understand completely, and neither will most of those who read this book. What I hope to do in this chapter is provide a template for how we can revitalize the family farm, rural America, and offer hope to those who love agriculture as much as I do, and in so doing, return freedom and agency not only to consumers, but also to individuals who want nothing more than to work hard and produce food.

We are living in a transition period, a moment in history in which each individual must decide what type of future they want to see in America. As citizens of this country, we must ask ourselves whether we want the independent family farmer to survive. Philosophically speaking, is this demographic necessary to the idea of America? If the family farmer can no longer compete in the marketplace, should these people and this way of life die out? Should they become the price of progress in the 21st century, just as the Lakota were in the 19th century? Should they give way to Big Ag, consolidation, synthetic protein, and Tech-controlled food production? There are individuals who, because of their massive wealth and influence, are attempting, with some degree of success, to make this vision a reality. However, each one of us, every person reading this book, still has the agency to check the powers-at-be, to prevent a cooption of our freedom and self-determination, and ultimately save the independent food-producer.

Every time we buy something, we are voting.

Whether it is food, farm equipment, handbags, beer, diapers, toilet paper, or tires, we are voting to propagate that industry, that company, and those employees, shareholders, and investors working within. Do we ever sit down and think about this? Do we think about the ramifications a paper towel purchase could have on forests, wetlands, and wildlife? Do we research the carbon footprint of every company for each product we are going to buy at the grocery store or order online? I certainly don't. Instead, we look at price, name brands, and wait for sales. We might solicit advice from friends, family members, and coworkers about products they use and their experiences. We tap Facebook and social media influencers for advice on what products to buy and how best to live our lives. And maybe there isn't too much harm in this. An errant T-shirt purchase from a company employing child labor in Bangladesh can be overlooked. We will feel bad and hope the company goes under. As for the T-shirt, if it is cute and our favorite influencer is wearing it, then it will become part of our wardrobe. This is something for which we can forgive ourselves. But when it comes to food production, the stakes are much higher. We can live without the cutest, trendiest T-shirt on TikTok, as unimaginable as that might be, but we can't live without protein. It would be quite difficult to control a population through apparel. But every population can and will be controlled if their food source is coopted.

The synthetic protein movement has garnered lots of media attention, droves of investors, and billions in seed money. These companies make a variety of promises, but all of them claim that their products are superior to real meat and real protein. In the interest of full disclosure, I have never sampled any of these products. I doubt that I ever will. What these companies fail to realize is that proving the value of their offerings compared to the real thing is beside the point. There is much more at stake than protein. They are winning one battle at a time while the rest of us are losing the war. Let's say, hypothetically, that one of these companies produces a product that *does* taste better than the real thing. And not only that, but it is also cheaper, healthier, keeps longer, cooks easier, is better for the environment, and more accessible to all socio-economic classes. You can order these products from Amazon and have them at your door within hours, delivered by a drone free-of-charge, while you are on Facebook reading about your friends' most recent experience with said products that were ordered from Amazon and delivered to their homes by a drone free-of-charge. Let's say that all of the company's claims end up being completely true. This is progress, right?

We are entering a period when government policies and regulations will be increasingly influenced and controlled by Tech. Private citizens with money and influence will be setting the policies and regulations in this country as they see fit in order to fulfill their own visions of what is "best for us" and "best for the environment." One wealthy citizen's utopian dream could be a dystopian nightmare for you.

Bill Gates has entered the agriculture industry. He

is the largest private owner of farmland in this country, with roughly 250,000 acres in holdings currently. He is also heavily invested in the synthetic protein space. He envisions a world where all "rich" countries are devoid of livestock and animal agriculture. He believes this is one of the primary ways to curb our carbon emissions and save the planet. I read Gates' book, *How to Avoid a Climate Disaster: The Solutions We Have and the Breakthroughs We Need.* I thought he came across as a well-intentioned philanthropist who is doing more, by himself, than some entire countries to fight climate change. And his work prior to climate change in addressing the healthcare and food insecurity issues for the global poor through The Gates Foundation is noble and awe-inspiring. Gates is easy to like, and I found myself rooting for him and his work throughout most of the book. However, like any individual, he has blind spots that cannot be overlooked if we are going to have an honest conversation about food sovereignty. Gates writes:

> "The upshot of all this is that we'll soon need to produce 70 percent more food while simultaneously cutting down on emissions and moving toward eliminating them altogether. It'll take a lot of new ideas, including new ways to fertilize plants, raise livestock, and waste less food, and people in rich countries will need to change some habits—we'll have to eat less meat, for instance..."[3]

I agree with some of this passage, especially our need to waste less food, and Gates seems to be blazing the right trail. However, he neglects to discuss the impact this will have on people working in agriculture, consumer agency, and choice in the marketplace. Gates is also unable or unwilling to envision what will happen if his desires concerning animal agriculture in the US come to fruition. He mentions "[eating] less meat" and "[changing habits]," but he does not discuss what this will mean for the American family farmer, rural economies, and private land ownership. In an interview promoting his book, Gates goes even further:

> *"So no, I don't think the poorest 80 countries will be eating synthetic meat. I do think all rich countries should move to 100% synthetic beef. You can get used to the taste difference, and the claim is that they are going to make it taste even better over time. Eventually, that green premium is modest enough that you can sort of change [the behavior] of people or use regulation to totally shift the demand."*[4]

We have to think critically about the last part of his comments. What will "[using] regulation to totally shift the demand" look like? Who will be influencing these regulations? What will they consist of? How will they affect farmers, hunters, gardeners, and fishermen? What impact will they have on consumer rights, consumer agency, and the individual's right to food security?

98% of Americans are not involved in production agriculture, and 96% are not involved in hunting. However, these enormous demographics will be setting farming and hunting policies in the future. These policies are set, more and more, at the ballot box. Bill Gates and others are using books, social media, and social media influencers to persuade this demographic to vote accordingly. The rights of individuals to hunt, fish, farm, and garden will be eliminated. This will all happen in the name of progress.

We must remember that the easiest way to control a population is to control their food source. Do we really want Gates and others like him telling us what we can eat, when we can eat it, and how much we are going to pay for it? Do we want our food security placed in the hands of wealthy and influential private citizens who are heavily invested in companies that produce synthetic protein for our sustenance? Should this be our only option? Should we accept the eventual criminalization of all marketplace competition to synthetic protein in the name of saving the planet? Should any individual be able to mandate that consumers purchase protein ONLY from the industry in which he or she is heavily invested?

Bill Gates and other technocentric mavens are hopelessly committed to the idea that technology equals salvation in every aspect of society. Synthetic protein is the new Eucharist, and we should all become believers, baptized into the New World Order. If Jesus Christ were alive today, he would probably work for Microsoft. But not everyone desires a techno-utopia. Not everyone wants to work for Apple, or Facebook, or Amazon. Not all of us dream of a work-free world, AI, the metaverse, and social media-curated realities. There is still a demographic that desires to work hard with their hands, outside with animals and machinery and nature. This demographic will not go quickly or quietly.

We must ensure that there is a sector for these people to work in going forward. We must let the marketplace and consumers decide if there is a demand for real beef, real chicken, real eggs, real milk, real lamb, real turkey, real wool, real leather, and real fur. If what Gates and others in the synthetic protein space are offering is so beneficial to mankind, then there will be no need to change regulations or influence policy. Those of us laboring in the trenches producing the real thing will wake up one day and find that we are out of a job. Some will capitulate and work in tech, some will turn to addiction, and some will become a statistic. But the free market will have spoken.

No matter what ultimately happens, I will try, to the best of my abilities, to opt out of the real or synthetic corporate food system and continue to hunt, fish, farm, and garden for as many of my food needs as possible. I am fortunate enough to have that option. But what about those citizens who are unable to do this, but who are just as concerned about these issues as I am? What about those individuals who don't have the time or the land space, who might be elderly or disabled or have special needs? How are these people going to provide for themselves? Won't they have to be a part of some system? Won't they have to buy food from someone or some company?

If only there was a demographic in this country that wanted nothing more than to grow produce, raise cattle, tend chickens, fatten hogs, and finish lambs in a regenerative, pasture-based system. If only we had people who still wanted to work on the land, long days without getting rich, satisfied to produce nutritious food for their fellow citizens and happiest working with animals, on their own, in a rural area, surrounded by fields and forests and wildlife. If only such a demographic still existed, then those citizens who could not provide for themselves might be able to patronize a system that valued every component in the supply chain. A system that valued the land, the animals, the farmers, the truck drivers, the warehouse employees, the products, and the consumers. What might it look like? Are there still people in this country that want to work for something other than wealth, status, or ego?

We are being convinced one social media post at a time that synthetic protein is the best solution for our future and for the health of the planet. Investors will receive greater and greater dividends, consumers will have increasingly cheaper and more convenient "options," and the archaic tradition of farming will finally be eradicated from this country, relegated to history as an "unenlightened" period when people worked hard to produce real food for their fellow citizens, all the while engaged in endless drudgery, animal cruelty, and degradation of the environment. The time is close at hand, and we should be grateful. Eventually, we will eat only what we are told to eat, buy only what we are told to buy, read only what we are told to read, and believe only what we are told to believe.

It has often been said that young people are lazy, don't want to work, and would prefer things handed to them. After all, they were coddled growing up, told they were all champions and rock stars and that any sports team, school, or employer would be lucky to have them. They are members of a great generation and will do great things for humanity through innovation, software development, and automation, all while not working too hard or experiencing too much discomfort. Life should be easy and fun and pleasurable, after all. There may be individuals out there who believe this and try to live according to this "code." But I do know, however, that there are also people in this country who would like nothing more than to work hard on a farm producing high-quality, nutritious food for their fellow citizens. They would work the long days, make the sacrifices, and deal with the stress and uncertainty and demands inherent in any small business. When I was farming full-time, I worked sixteen to eighteen hours per day, six to seven days a week during the growing season. It was a relentless grind, even punishing at times. However, at the end of every grueling day, I would walk back to the house feeling a sense of fulfillment, satisfaction, and tranquility that I have been searching for ever since.

I believe strongly in the old saying that if you are looking for a helping hand, you will find it at the end of your arm. I also believe in a competitive marketplace, an entrepreneurial spirit, and a level playing field. Running a small business can be overwhelming and not everyone is cut out for it. Farming is no different. It is subject to the same economic realities and brutalities as any other small business. Not everyone who wants to be a farmer can be. Neither can everyone who wants to be a baseball player or a pilot or a teacher or a police officer. What we need more than a hand-out or more government subsidies or more suicide hotlines—as important as these things are—is to spread opportunity for the family farmer to succeed in the 21st century. Tech and Big Ag are holding all the cards, have all the lobbyists and politicians, and are slowly choking the life out of the family farmer, one bankruptcy, foreclosure, and suicide at a time.

I am going to pretend for a moment that I have Bill Gates' wealth and influence. I have Bill Gates' power. How would I help my fellow citizens? What is my vision? If I were Bill Gates or any other ultra-wealthy citizen, I would create a *private* subsidy program, independent of any government funding. I would subsidize all of those individuals who cannot hunt, fish, farm, or garden for themselves to instead patronize independent family farms across the country for their produce, protein, and other food needs. I would help the single mother in downtown Saint Louis who is working three jobs and raising four kids in government housing in a food desert, simply trying to make it through each day. How could she participate in this system? How could she possibly afford to feed her children food that didn't come from Walmart or Save-A-Lot or Aldi's or Amazon? How could she use her food dollar to vote for the independent food producer? This is where I would come in. I would provide this person with vouchers to cost-share her purchases of real food from local farmers in her community. She would have options when deciding what to feed her children. She wouldn't be relegated to the corporate food system that respects nothing but the bottom line and certainly does not respect her. These products could be delivered to her door with instructions on how to cook and prepare meals. She would feel proud and empowered that she was participating in an alternative food system, a real system benefitting real people and real communities, and at the same time consuming the highest-quality nutrition available. Small farmers could compete for her business in a fair and transparent marketplace. Rural areas would be revitalized. More children would be able to return home to work on the family farm alongside their parents and grandparents. They would marry and raise their children in the communities they grew up in. Their tax dollars would stay local, benefitting the school systems, the roads, bridges, and other infrastructure. Hospitals would be built, and with any luck, mental health professionals in rural areas would take up golf to kill time while they waited for a dwindling supply of patients to make appointments.

Bill Gates and his ultra-wealthy friends could pool their money together to create an investment fund that would provide capital for this subsidy program in perpetuity. This system would benefit and respect people at every step in the supply chain. Farmer suicides and food deserts and nutritional deficiencies amongst the poor would become a thing of the past. Consumers would eat real food produced by real people on farms working hard every day in an occupation that provided fulfillment in life. People could spend the prime years of their lives involved in meaningful work, building something that will last, instead of spending countless jealous hours on Facebook reading about others who are involved in meaningful work building something that will last. That's what I would do if I were Bill Gates.

Of course, the real Bill Gates isn't going to do that. He is going to double-down on synthetic protein technology and investment. His answer to the epic consolidation in Big Ag is epic consolidation in the Synthetic Protein space. He is going to purchase more farmland to add to his vast holdings. The rich in Silicon Valley will get richer, all the while telling us that they are making our lives easier and doing what's best for us and saving the planet. This is progress, it's a beautiful thing, and we should be grateful.

As for me, I will watch the spectacle with tired eyes. Then I will get back to hunting, fishing, farming, and gardening until the day comes when Bill Gates shows up at my house and knocks on my door and asks to purchase the last tract of land in America still producing real food by a real person living a real life. A life full of hard work, uncertainty, and freedom.

IX

BEES IN THE HIVE

*His sudden, offstage ending left him suspended forever between East and West, past and future, to be misremembered as needed by each new generation.—**T.J. Stiles, Custer's Trials***

He couldn't wait any longer. The greatest opportunity of his life was slipping away right before his eyes. He should have expected as much from Benteen. Ever since the Washita, Benteen had been seething with revenge, waiting for his chance. He never should have divided the command. That was a mistake to be sure. But he could still win with a little luck and hard fighting. Every second he waited for Benteen, however, the warriors grew in number. If he waited much longer, a final and decisive thrust north into the heart of the village would be impossible. To hell with Benteen, the insolent bastard, he had to attack, and he had to attack right now.

Lieutenant Colonel George Armstrong Custer's body was found, naked and partially mutilated, by members of the Dakota Column two days after he was killed. His ears were reamed out with an awl, and an arrow had been thrust into his penis. One of the most virile, aggressive, and successful cavalry commanders in the country would be rendered impotent in the afterlife. He would, however, be etched into American History as a Civil War Hero, Legendary Indian Fighter, Darling of Americana, Martyr, Villain, Fool, and Scapegoat. How could he be seen in such contradictory terms? Why is Custer's legacy so complicated and so difficult, even today, to distill into a simple and acceptable narrative?

Some of the answers may be found in Custer's character. Long before his death in Montana, Custer had already lived many disparate lives. He was a gallant Civil War hero adored by his men, a peacetime martinet who exacted brutal and even illegal discipline on his command, a failed politician and entrepreneur, a doting husband and prolific philanderer, a competent marksman and hunter who accidentally shot and killed his own horse during a buffalo chase. Custer was a man full of contradictions. As T.J. Stiles writes:

> *"He was complex. He could be emotionally sensitive. He had taken pity on enemy civilians, dead and wounded comrades, and former slaves whose backs told the history of whipping in keloid Braille. He loved the theatre, which is the mimetic experience of others' lives. Yet he was a practical joker, finding amusement in causing embarrassment or pain. He had a faculty for sympathy, but it had a faulty switch. When it malfunctioned, he saw no farther than himself. It may be why he was so good at killing."*[1]

During the Civil War, Custer led cavalry charges from the front, often galloping full-speed with saber drawn as balls and shells cut down men all around him. His gallantry and consistent success propelled him to the brevet rank of Brigadier General at twenty-three. However, he dressed flamboyantly, grew his blond hair long into locks, brushed his teeth after every meal, and obsessively washed his hands. He never drank and rarely swore, but he was a compulsive gambler and always in debt. He was an animal lover and often brought more than a dozen dogs with him while on campaign. But he had no compunction in pulling the trigger on any fowl, buffalo, bear, or antelope that wandered into range, whether he needed the meat or not. He claimed to respect native cultures and often participated with his scouts in their own rituals, even going so far as to make corrections if an element was forgotten or poorly executed. Inexplicably, he desecrated the scaffolds and burial sites of warriors killed in battle. Despite holding the tactical advantage, he also refused to attack a Cheyenne village in order to negotiate for the lives of two white women who had been tortured and raped. However, he allowed his own men to take captive Cheyenne women and girls as bed partners, with or without their consent.

Custer's most valuable qualities—aggression, lucidity under intense fire, high battlefield IQ, and relentless will to win—served him and his superiors quite well at Yellow Tavern, Cedar Creek, Third Winchester, and Gettysburg. But the Civil War was over. Custer's natural talent for warfare was neutralized by his inability to succeed at anything else. Shortly after witnessing the abject slaughter at Antietam, Custer wrote, "...I must say that I shall regret to see the war end. I would be willing, yes glad, to see a battle every day of my life."[2] 3,600 men died, and 17,000 were wounded during the bloodiest day in American history.[3] Custer was not sickened or traumatized or horrified. Instead, he was enthralled as if he had stumbled across his life's calling. His subsequent actions during the rest of the war proved that his written words were not hyperbolic, and his reputation as a Union hero was sanctified in a crucible of blood, steel, horses, and victory. But the war could not last forever and, unfortunately for Custer, he would eventually find himself an officer in a peacetime army with a drastic reduction in rank, necessitating a transition from warrior to manager—a transition he was ill-equipped to make.

Growing frustrated and bored with the post-war army, Custer attempted a political life that was disastrous and did nothing but increase his enemies. He even alienated a close friend and mentor, General Philip Sheridan, who had shepherded him through his meteoric rise in the ranks. Multiple failed business ventures including real estate, horse trading, silver mining, and the stock market, further impoverished him. He was fond of women, slept with Indian captives, and his marriage was chronically riddled with rumors of infidelity on both sides. He ultimately deserted his command while on campaign in hostile territory, to attend to his troubled union, and was summarily court-martialed and suspended without pay for one year.

Custer had been adored by his men and the nation during his epic Civil War career. His post-war track record was much more mixed, however, and he had been in disgrace financially and professionally before being assigned to the Sioux Campaign. He was hoping to revive his political career and speaking tour prospects with one last great victory. His Civil War record spoke for itself, and with a legendary battle to cap his Indian-fighting career, he would once again become a gallant and dashing American hero. But he needed to do more than just capture Indians or negotiate a surrender. He needed to win a decisive and glorious battle in what he had already decided would be his final campaign. There was hardly an officer in the field who was more motivated politically, financially, and personally to attack Indians than Custer. This victory would restore his reputation, his bank account, and his future.

When it came time to select a commander to lead what was expected to be the terminal offensive against the Lakota in the spring of 1876, Custer was chosen. But his political maneuvering and self-promotion chaffed President Grant, who ultimately removed him from the campaign. Grant decided to give General Alfred Terry overall command instead. Terry was not an Indian fighter and generally preferred the comforts of a desk over a saddle. He requested that Custer come along to serve under him and lead the 7th Calvary in the field. General Sheridan also appreciated Custer's unique experience and joined Terry in lobbying Grant for a reinstatement, albeit in a lesser role. The U.S. Army did not have a deep bench of reliable and consistent officers. What the top brass needed was an "old school," experienced cavalry commander, one who *would* attack, given the chance, and would produce concrete results. The Sioux War was dragging on too long, and the American public had grown weary with the handwringing and appeasement of the Indians. Grant, Sheridan, and Terry needed an officer that would find the village, attack en masse, and break the back of the Lakota resistance. Despite Grant's reservations over Custer's politics, he knew Custer possessed the singular qualities the mission demanded. Sitting Bull and Crazy Horse just weren't getting the message, and it was time to take off the gloves.

The problem throughout most of Custer's career in the West had been finding Indians in numbers large enough to justify an attack. Most of the officers in the field and in Washington agreed that locating Sitting Bull's village and attacking before a mass exodus transpired was the most prominent concern. Like many of his fellow officers, Custer had spent an inordinate amount of time tracking Indians, only to see a once-promising trail evaporate into a labyrinth of sub-trails, each smaller than the one from which it came. It is staggering to think of the amount of horse flesh, grain, and sanity that were lost chasing phantom Indians on the open plains between the end of the Civil War and the Battle of the Little Bighorn. The conventional wisdom, therefore, compelled Custer to make haste and strike while there was still a battle to be had.

When Custer's scouts finally located a very large trail near Rosebud Creek in present-day Montana, Custer was elated. Custer's subordinate, Major Marcus Reno, had already conducted a valuable reconnaissance mission that all but confirmed Sitting Bull's village was in the vicinity. This trail likely led to his camp. However, the scouts quickly became concerned when they realized that this trail was not dissipating but was becoming one that was "more than a mile wide, the earth so furrowed by thousands of travois poles that it resembled a plowed field."[4] They eventually found the Indians' massive pony herd and implored Custer to look for himself while on top of the Crow's Nest. Even with the aid of field glasses, Custer struggled to see what they had seen: fifteen to twenty thousand ponies grazing in the valley near the village, fourteen miles away. Whether he actually could not see the ponies from this perch, or his brain was still operating under the errant assumption that there simply could not be that many Indians gathered in one place, can never be known. After this dubious reconnaissance, Custer decided an attack was in order. The scouts advised him not to divide his command in the face of such a large force. Custer ignored them. Captain Benteen, for his part, also recommended that Custer keep the command together. Custer dismissed him as well. Custer was still hopelessly committed to the belief that the biggest tactical problem would be preventing the Indians from escaping and scattering into the Bighorn Mountains to the south. He was so concerned with losing the opportunity for a grand battle that he never imagined the Indians would not only hold their ground, but would counterattack en masse. He eventually divided his roughly six-hundred-strong command into four units under Major Reno, Captain Benteen, himself, and Captain McDougal, assigned to guard the pack train. Dividing a command was a common battlefield tactic in Indian warfare, and Custer had employed it successfully in past battles. On this day, however, it was a fatal mistake. Upon seeing the village in its totality and finally understanding what he was up against, Custer urgently sent for Benteen to rejoin him, effectively admitting that Benteen and the scouts had been right all along.

Custer's plan had been for Benteen to swing far south of the village to prevent any escape in that direction towards the Bighorn Mountains, which Custer deduced would be Sitting Bull's most likely course of action. Reno was to attack the village from the south as well while Custer attacked from the east. Reno charged the southern end of the village as ordered—the opening act of the battle. He faced stiff and unexpected resistance from several hundred warriors. After sustaining only a single casualty, he halted his men a few hundred yards from the village. After a period of indecision, contradictory orders, and at best mediocre leadership, he ordered his men to retreat into a patch of timber in an effort to reorganize and prepare a makeshift defensive position. He was conferring with Bloody Knife, Custer's favorite scout, to determine what the Lakotas' most likely course of action would be now that the command was hunkered down in the timber. At that moment, Bloody Knife was shot in the head, the contents of which splattered all over Reno. That was the last straw as far as Reno was concerned. He told any soldiers within earshot that he was heading for high ground, and they could follow him if they wanted to. Reno effectively ordered an every-man-for-himself retreat that grossly violated all standard military doctrine for withdrawing a unit under fire. The soldiers who were unable to hear Reno's orders over the deafening roar of gunfire were left to defend themselves as best they could. The Lakota siezed upon this opportunity and charged into the retreating soldiers' lines as they left the timber, shooting them at point-blank range, pulling them off their horses, and killing them with a chilling efficiency that spoke to their superior horsemanship and close-combat proficiency. Reno survived and was now on top of the hill that would bear his name, licking his wounds, partially drunk and getting more so, and completely demoralized. Half of his command was dead, wounded, or missing. Custer had promised to support him but was nowhere to be found. Disgusted with his mission and coming across nothing, Benteen returned to find Reno and his battered command. Shortly thereafter, he received Custer's final and desperate message to "Come on." If Benteen had hurried to Custer as ordered, the entirety of his own command might have been annihilated as well. Or perhaps the addition of his one hundred thirty relatively fresh troops, plus the remnants of Reno's command, would have been enough to turn the fight into a siege, similar to what eventually happened on Reno Hill during the second half of the battle. Many soldiers still would have been killed, but probably not an entire command. And maybe not Custer.

Benteen's valiant efforts to save the rest of the Seventh Cavalry during the second day of the battle have, in many ways, given him a pass. For all his faults, Benteen was consistently cool under fire. As the Lakota surrounded the surviving soldiers on Reno Hill, Benteen provided a rallying symbol for the men, not so much for the substance of his orders, but for his willingness to lead upright and exposed in the lines as the bullets were flying. In addition, he wisely recommended that Custer keep the command together, advice Custer should have heeded. Reno has received the lion's share of criticism for his failed attack, pell-mell retreat, and drunken abdication of leadership. But Benteen delayed for over an hour after receiving his orders before moving towards Custer. By then, Custer's command was all but finished. They were facing a counterattack that outnumbered them at least six to one, not to mention many of the Indians were carrying repeating rifles, a significant advantage over the soldier's single-shot carbines. Crazy Horse, Gall, Crow King, Lame White Man, and others were fighting with such ferocity and skill that a Blackfoot named Kill Eagle likened the Lakota counterattack to "bees swarming from a hive."[5] After discovering that his family had been killed early in the battle, Gall, for his part, chose to fight with nothing more than a hatchet.

He mounted his horse, rode into the fight, and hacked his way through Custer's command "like a wolf through a flock of sheep."[6]

Custer was now lying three miles away from Reno, Benteen, and the rest of the command, a victim of questionable performances on the part of his subordinates, but also a victim of his own constitution. He was an increasingly desperate man who needed a blood-soaked victory. Everyone knew the Indian Wars were almost over. Therefore, Custer did his best to outpace the other units involved in the Sioux campaign to secure the entirety of the glory for himself and his command.

Custer was an instrument employed by the United States Government in a much larger campaign of oppression. Throughout his career, he had fought hard and valiantly. As his command was disintegrating at the Little Bighorn after the aborted attack on the village at Medicine Tail Coulee, Custer still attempted to attack the village from the north, the only chance to check the overwhelming force against him. If he had succeeded and breached the village, he could have taken women and children hostage and perhaps neutralized the Indian's counterattack. This, of course failed, but what is telling is that Custer did not leave the battlefield, did not even try as far as we know. He may have been mortally wounded early in the battle, or he may have fought to the end, one of the last soldiers to die.

Custer was fighting as much for himself as for his fellow men and his country. He needed this victory, a victory that would ensure a lucrative speaking tour on the lecture circuit and a dark horse run as the Democratic candidate for president. It is fair to say he was selfish, reckless with the lives of his men, and foolish for rejecting the advice of his scouts not to divide the command—scouts that he had previously praised for their competence and loyalty.

When analyzing Custer's performance during the final act of his life, it is tempting to lay the vast majority of blame at his feet. A more cautious or traditional cavalry officer would not have attacked, would have waited for the Dakota Column, all while surveilling the Indians and their movements. Custer is remembered for his epic failure, a failure of imagination, of character, and ultimately of sound military judgment. He got a lot of men killed needlessly on both sides of the battle. In the final analysis, it is tempting to declare that Custer was a deeply troubled man, struggled in almost everything he attempted in life, and only achieved any success in the Army because he happened to be in the right place at the right time.

We need our heroes and our villains to fit neatly into each category. Custer is no different. He was a racist, a philanderer, a degenerate gambler, an opportunist, a dandy, a narcissist, and all in all, a man of low character, best left relegated as an obscure footnote in our nation's history. He did nothing to benefit any of us living today, and his legend and reputation are an embarrassment, one more example of a man benefitting from white privilege, whose ultimate station in life far exceeded his abilities and accomplishments.

I will admit that my own beliefs about Custer largely fell into this camp before I studied his life. He was surely a one-dimensional person with few redeeming qualities, and only received lasting fame because he died in a Hollywood ending, a real-life script that studios salivate over. However, as I read more about Custer's service to our country, I began to learn that he might not fit so well into the villain category after all, and that there was a period in his life where his battlefield performance was nothing short of heroic. This didn't square with the narrative and would surely contradict what so many of us have been led to believe.

Custer's disposition as a tenacious, overconfident, and fearless combat commander is largely what sealed his fate in Montana. He had demonstrated sound judgment at times during his military career, actions that had saved his own life as well as the lives of his men, but his instinct was always to attack, to charge and carry the day through sheer aggression and will to win. This singular constitution would ultimately serve him well at one of the most pivotal moments in the Civil War.

More than one historian has run out of ink criticizing General Lee's decisions during the three-day bloodbath at Gettysburg. However, Lee's strategy came frighteningly close to fruition on the final day of the battle when the Southern units under George Pickett and J.E.B. Stuart nearly linked in a simultaneous attack that could have overrun the entire Union line and altered the course of the war. Custer played a critical role during Pickett's charge, one that is rarely mentioned when discussing his legacy:

> *"Unknown to Custer, Lee had decided to win the Battle of Gettysburg with a decisive thrust at the center of the Union line. He had called on General Stuart to take his much-admired horsemen and swing around the Union right flank in a solid mass. He was to get into the rear of the Union infantry during the frontal assault and pursue and destroy the retreating foe. The Michigan Brigade was the last obstacle in the way."*[7]

A Southern cavalry unit led by Wade Hampton, serving under Stuart, was seeking to destroy the artillery guarded by Custer's men and advance to the rear of the Union line, ideally in concert with Pickett's charge to the front. Lee strategized that the combined force would crush the enemy and force a retreat, ending the battle of Gettysburg and sealing victory for the Army of Northern Virginia. Hampton's unit fanned out and advanced towards Custer. Custer understood clearly what the Union would be risking if Hampton overran his position. He was at the apex of his Civil War career. Despite facing a capable and experienced enemy and heavily outnumbered five to one, Custer charged:

> *"The Union cavalry suffered roughly 250 causalities in this fight, nearly 90 percent in the Michigan Brigade. Gregg had commanded, and commanded well, but Custer had led, in person and in the place of crisis. Somehow he survived without a wound, despite his plunge into the heaviest fighting.... Most important, Custer was not reckless with the lives of his soldiers, however much he was with his own. Though an*

*aggressive commander, he had not attacked blindly. He made
the most of the new technology of the repeating Spencer rifle
by deploying many of his soldiers on foot, and he put his well-
manned artillery to good use. He chose the right moments
to charge, essentially counterpunching after his opponent
committed himself."*[8]

Custer's singular abilities and leadership that helped
carry the day at Gettysburg when the fate of the Union
and the war was in doubt, when Pickett was advancing
with twelve thousand men bearing down on the Union
front, while Stuart was simultaneously attempting to
reach the rear, were exactly the same attributes that
compelled him to attack an enemy of unknown strength
with little reconnaissance on June 25th, 1876. We can
criticize Custer for his failures at Little Bighorn, but
these liabilities were crucial assets at Gettysburg, and it
will forever be debated what might have happened had
a more "cautious," "reasonable," or "traditional" cavalry
officer been leading the Michigan Brigade against
Hampton's command.

Custer's performance at Gettysburg was not a
fluke or a "one-off." He did not simply get lucky against
an inexperienced or outnumbered enemy. To the
contrary, he defeated Wade Hampton, arguably the
most competent cavalry commander in the Confederate
Army. He repeated this performance many times during
the war, albeit in less pivotal battles and against less
formidable opponents. All in all, he fought in a total
of eighteen engagements and had eight horses shot out
from under him.

Ever in search of the limelight, Custer was present during Lee's surrender at Appomattox. General Sheridan managed to acquire the very table upon which Grant drafted the terms of surrender. He sent it to Libby Custer as a gift with an accompanying note:

> *"My dear Madam—I respectfully present to you the small writing table on which the conditions for the surrender of the Confederate Army of Northern Virginia were written by Lt. General Grant—and permit me to say, Madam, that there is scarcely an individual in our service who has contributed more to bring this about than your very gallant husband."*[9]

Unlike Custer's reputation as a legendary Indian fighter, this commendation could be supported with plenty of hard evidence and concrete results. Robert Utley writes:

> *"The surrender table acknowledged and symbolized one of the Civil War's most extraordinary careers. From Gettysburg to Appomattox George Armstrong Custer wrote a virtually faultless record of battlefield success, made the more remarkable by his sudden rise and his extreme youth. His generalship combined audacity, courage, leadership, judgment, composure, and an uncanny instinct for the critical moment and the action it demanded. He pressed the enemy closely and doggedly, charged at the right moment, held fast at the right moment, fell back at the right moment, deployed his units with skill, and applied personal leadership where and when most needed."*[10]

It was very disconcerting for me, and perhaps other readers as well, to learn that the freedoms so many Americans enjoy today are, in some small part, due to the exemplary combat performance and steady leadership not only at Gettysburg, but throughout the Civil War, by the most vilified calvary officer in our nation's history.

***

The Battle of the Little Bighorn was a military disaster of epic proportions, one of the worst in the history of our country. And most of us learned about this in school. But what do we remember? What were we taught? That Custer was an arrogant monomaniac who foolishly divided his command before attacking an enemy of unknown strength, and ultimately got what he had coming? And wouldn't it be so nice if we could accept that—if we could take as Gospel the two inches of text reserved for the battle in our grade school history books.

The whole truth, as is often the case, is much more complicated. We have to step back and ask what was Custer doing in Montana in the first place? After all, he was not a private citizen who had financed and trained his own command independently. Why was the Seventh Cavalry dispatched to kill Indians? Who would benefit from this campaign, if successful? What did the Indians represent, and why were they considered so dangerous?

Custer came of age when America was transitioning from an agrarian society into an industrial powerhouse. With the War between the States settled, it was time to return to the dormant policy of American Progress. The railroads had to be laid, the West had to be settled, and gold and minerals in the Black Hills had to be exploited.

And the Indians were in the way. They were killing settlers, farmers, and miners. Maybe a few treaties had been retracted, and maybe they did have a right to be a little angry, but these things happen. And we really needed the Black Hills. We even offered the Indians a fair price, but they wouldn't sell. Well, we tried. ***If only they could see what was best for ~~us~~ them***. America was becoming industrialized and simply couldn't tolerate stone age cultures doing what they pleased inside its borders. This was progress, and yes, there is always a price to be paid, just not by us. Those that refused to capitulate would be made to do so by force.

The average man or woman in the street in 19th century America accepted the fact that the Indians had to be put on reservations. If a few had to be killed on the field of battle to facilitate this process, so be it. The Grant administration had already tried, with mixed results, to entice the Indians to accept reservation life and the benevolence of Washington. They would receive regular rations, medical care, and lodging. They would find stable employment as farmers, policemen, and military scouts. They would worship a Christian God and send their children to Christian schools. The Lakota language and culture would be banned and eventually lost to history. The Indians would eat what they were told to eat, buy what they were told to buy, read what they were told to read, and believe what they were told to believe. If they adhered to these pronouncements and behaved well, they could retain some firearms and hunting rights. This is what the government was selling, but not enough of the Indians were buying. Killing a way of life was proving to be very bloody, expensive, and lengthy. The Treasury Department was spending one million dollars for every warrior killed on the battlefield. Such a costly war was untenable. The fundamental problem was that the Indians were just too resourceful. Many of them didn't need Washington, at least not yet. Their desire to provide for themselves was infuriating, and attempts to create dependencies through alcohol, tobacco, coffee, and sugar met with some success, but were still insufficient. The government had to devise a plan that would ultimately force the Indians to accept reservation life as their new reality.

It was generally understood that the demise of the buffalo would do more to enslave the tribes than anything the Army could do. If the buffalo were eliminated, the Indians would lose their primary food source, tools, clothing, lodging material, etc. European trade goods had already replaced many of their daily essentials, but there was always a fall-back. In times of scarcity, a warrior could still strike out on his own, kill a buffalo, and provide for his family indefinitely. Not only did the buffalo have to be exterminated, but more importantly, what the buffalo represented. The entire cultural thought process of doing for yourself, of working for your food, procuring what you needed through your own efforts with a skill set passed down for generations—this had to be eliminated. The most crucial transformation the government desperately needed the Indians to adopt was the abdication of their autonomy and its hallmark of resourcefulness in exchange for providence and its hallmark of enslavement. More than Crazy Horse or Sitting Bull or any one man in battle, this belief system had to die.

Custer's defeat was a severe setback, but only temporary. After all, the railroads were being laid, the buffalo hunters were increasing their slaughter, and the Army would eventually commence a dogged and sustained post-Bighorn campaign to round up the remaining "hostiles." The US government knew that the easiest and most effective way to exterminate a way of life and control a population was not to kill men on the battlefield, but to first eliminate and then control their food source. Many Indians had readily accepted rations and were making the transition, often arduously, to Christian farmers. However, the traditionalists refused to do what was best for them. Exterminating the buffalo left them no choice. Eventually, even Sitting Bull, Crazy Horse, and their followers were starving. Both ultimately surrendered to US authorities. And it is no coincidence that both were subsequently killed on reservations, under the auspices of the very government that promised them food, shelter, income, and medicine, in perpetuity, if only they would give up their culture and way of life.

Custer had been dispatched as a catalyst to force the native capitulation to Washington. His mission was seen as a righteous crusade for progress. It was time to turn the page in American history, and Custer's anticipated victory would usurp the Indians' agency once and for all. His failure was a catastrophic blow to American pride and American confidence. Did Custer have it coming? Maybe. Did the nation have it coming? Absolutely. Custer was a product of the nation, sent by the nation, to do the nation's bidding. His failure was America's failure. Any blame he received then and now is short-sighted and misses the larger picture. His tactics will forever be debated, but the mission had always been quite clear: Become the instrument of progress, channel the power of the state, and eliminate a way of life.

Custer was a metaphor for America at the time. The country was transitioning from the honor and pride of the antebellum era, largely achieved through slavery, to the technological progress and development of the late 19th century. Once the Indians were safely confined to reservations, true progress could take place, and the country could catapult itself into a land of opportunity for all who accept the New World Order and embrace the tenets of progress, progress, progress, technology and progress and more technology and more progress, costs be damned. Custer was the harbinger, the ambassador of the New America, whether he fully understood his role or not. He was the poster child for what happens to unbridled imperialism, the fall guy for all of America's post-colonial sins. He will likely always be seen as a hero, fool, martyr, or villain, depending on the lens. I believe he died on that hill in Montana far from a hero, but more than a fool. He died a soldier following orders, and he died a human being.

Custer's legacy is subliminal, layered, and complex.

As a country, we have always wanted to put him and his battle in a nice little box of failed leadership, recklessness, and arrogance attributed to one person. We are not encouraged to look any closer because if we do, we might find a troubling history. We might find that Custer was not exactly who we need him to be, and his state-sponsored mission demonstrates the obsession this country has with "progress" at all costs, one which still exists today. We might see that the conditions under which Custer was deployed are very similar to the conditions in which we find ourselves today.

The Battle of the Little Bighorn closed a chapter in American history, and a way of life ended. Custer may have been defeated, but it was the Lakota who lost in the long run. Freedom died that day, and it is dying now. History does repeat itself, and many Americans will see that the world they once knew is changing at breakneck speed. Another Little Bighorn is on the horizon, another chapter is poised to close. The New America will have little use for those of us who refuse to accept the benevolence of Tech, just as it did for the Lakota who refused to accept the benevolence of Washington.

Our obsession with technology has become our new way of life. Stop doing for yourself, stop working, stop thinking. Be on the right side of progress and join us in this new mission to transform the nation into a fully-automated, fully-curated, fully Tech-controlled reservation. This is the life you have been waiting for. This is the new American Dream. A dream that Tech is currently manifesting in our society with a magnitude, an efficiency, and a following that a thousand Custers and a thousand Little Bighorns could never hope to accomplish.

X

BRAVEST MAN I EVER SAW

If your world was dying, would you not fight? **—Pekka Hamalainen, Lakota America**

The wolves had moved into the next valley. He listened intently to their distant howling and breathed a sigh of relief. He was out of ammunition, and it was impossible to know when he could acquire more. He cut the elk's track half a mile to the south. The track was fresh, still soft at the edges, and he could smell urine. It was a cow traveling alone. She likely heard the wolves as well and was headed down into the timber. He watched the wind indicator he had tied to the tip of his bow. It was fashioned from an eagle feather, and it caught the slightest breeze. He saw it flutter, hesitate, and flutter again. The wind was in his favor.

He had not eaten in five days. The hunger pangs had subsided, and the nausea had dissipated. Now he merely felt an emptiness and a dull, but persistent ache. He thought of the women and children who were starving as well, many too weak to leave their lodges. The thought of their suffering kept him moving despite the cold. He desperately needed meat. A large cow elk could yield two hundred pounds. It wouldn't last long, maybe a week. But it might keep more from leaving and heading for the agency. But he would take anything at this point. He would draw on any living animal—porcupine, skunk, weasel—anything with flesh and blood.

The track was coming easier as the cow left the high country. Her hooves were burying deeper into the snow. It was so cold that he had to stop and make a fire to protect his hands and feet from frostbite. As he sat warming himself, he thought about his life and his people. He thought about his friends who had died alongside him in battle. What would they think now? How would they feel about their sacrifice? He was dressed in rags. His men were constantly outnumbered and short on ammunition and rifles. The children were sick with agency cough. He was losing more families every day. They were leaving because they were sick and hungry and desperate, but mostly because their hearts were broken.

He followed the track. The cow had slowed her pace. He paused several times to listen. It was clear she was looking to bed down. This was lucky. A bedded elk would give him time to close the distance. He was hypnotized by the thought of warm, blood-soaked meat only a few hundred yards away. He knelt beside the track, took off his deerskin pack, and placed his quiver beside it. The timber was becoming thicker up ahead, and he couldn't risk snagging a branch or catching some brush and alerting the cow. He needed to travel as light as possible. He selected two arrows and gripped them against the bow riser. He slid his fingers over the cedar shaft and turkey feather fletching. He checked his nock grooves to ensure they were free of debris. He lightly passed his thumb over the steel arrowhead and drew blood. He said a prayer and asked for help in his quest. He asked the elk for her life.

He waited as long as possible. It was so cold that his fingers were numb almost immediately after removing them from his buffalo robe. He wanted to give the cow more time, ideally waiting until she rose on her own before letting an arrow fly, but he couldn't sit in ambush that long. He was cold and starving and desperate. He began the stalk.

He couldn't see the elk, but the tracks led into a patch of timber two hundred yards below him. The timber was near a small stream which gave her access to water and shelter from the biting wind. It was a good place, and he often wondered why so many people died in these mountains. If they had cut an elk track, they would have found shelter, water, and meat. The wind was picking up, almost howling at this point. He moved when it blew and rested when it subsided. His moccasins emitted a slight crunch each time they sank into the snow. He feared the cow would sense his presence and flee upstream. He had slowly worked his way down the hillside and could see the timber a hundred yards ahead. An hour passed, and the wind was blowing through him, freezing his bones. He could hear the stream gurgling and choking its way under the ice. There was little cover. He clung to the hillside, hoping he would blend in if the cow could see in that direction. If she was facing him, his chances for success were minimal. Another hour passed. He was now at the edge of the timber. He could see the tracks right beside him, but it was difficult to tell how far into the timber she had traveled before she bedded.

It was getting thick, and the cow could be just yards away. He stopped and listened. The wind was still blowing hard but would subside intermittently to almost silence. He could hear nothing. Was she still in there? Had she left hours earlier? Was he stalking a ghost?

He scanned the timber, searching for any sign of the cow—an eye, an ear, a flash of brown, anything. A distant howling startled him, and he was reminded that there were other hunters in these mountains looking for meat. Then he saw movement and dropped to one knee. The cow was bedded fifty yards in front of him. She had heard the wolves as well and turned her head to locate the source. The cow listened intently for a minute or so in his direction, then turned her head back. He waited. He wanted to continue to wait until she rose on her own, but he was so cold that even minutes without moving was unbearable. The wind was increasing, and he could feel the temperature dropping. If he stopped moving too long, he would shiver uncontrollably and wouldn't be able to draw his bow. He covered ten yards without making a sound. The cow's breath was steaming out of her nostrils and into the air. Her eyes were bright and alert. He was now within forty yards of meat and hide that would buy him time, buy his family and his friends time to live on their own terms, free and healthy in the mountains, away from the Army and the forts and the agencies and the traders and the constant pull of all those telling him to give it up, to come in and let the old ways die. The cow was more than flesh, blood, and bone. The cow was freedom.

He carefully closed the gap to twenty yards. He dared not go further. He was standing behind a large aspen for cover, only exposing his forehead and his eyes. He had an unobstructed shooting window and a clear view of the cow. She was facing upstream. He slid an arrow along the riser and placed the nock on the string. He had no shot while the cow was bedded. He couldn't risk a bad hit or a wound or a miss. The thought of returning to camp empty-handed made him sick. He needed the cow to rise and expose her heart and lungs, but his fingers were quickly going numb on the string. He couldn't wait any longer. He had to kill her now. He breathed in deeply, tightened his grip on the riser, and came to full draw. Then he said a prayer and uttered a low guttural grunt. The cow quickly stood and looked in his direction, scanning for the "bear." She located him and froze. Her eyes widened, every muscle in her body tensed, and she was planting her hooves preparing to bolt. But just as the cow rose, he picked a spot behind her shoulder, aimed, and released the arrow without thought and without hesitation. He watched as the arrow arced through the air and buried deep inside her lungs up to the fletching. She whirled and crashed through the timber. He listened intently as he lost sight of her. Then he hurried over to the cow's bed to confirm the hit and saw pink, frothy blood in the snow. He knew the arrow had penetrated both lungs, and she would be dead in seconds. He nocked the other arrow and waited and listened. He heard a long grunt and then a crash, and he knew the cow was down. He waited. Then he glided along the blood trail like a fox in the dead of night until he saw the huge brown body in the snow. He crept closer and tossed a stone. No movement. He came upon the cow and touched his bow to her eye and saw no reaction. He knelt beside her and slit her throat. As the rich, dark blood flowed

through his hands, he placed his head on the cow's head and
wept. He felt the warmth of the cow and the softness of her
hide. He said a prayer and thanked the cow. Then he opened
the body cavity as quickly as possible and plunged his frozen
hands inside. When he regained feeling, he cut her heart out
and gorged himself. The blood flowed down his chin onto his
chest and soaked his robe. He stood up and looked at the
cow, marveled at how massive she was, maybe the largest he
had ever killed. He began the butchering process knowing she
would freeze quickly. As he packed the meat out, straining
from the hundred-pound loads, he welcomed the pain and the
exhaustion and the cold. He welcomed the burden. The cow
gave him another week of freedom, another week living with
his people, on his terms, in the mountains. With her warm
blood inside him, he felt reborn. From her death, he found life.

* * *

Custer was furious. He was scanning the village ahead
of him and watching the panic, the confusion, and the
fear. He had a short window to attack, and it was closing.
The warriors would quickly organize and defend the
village. There would be many more than he thought, but
he still held the element of surprise. If Benteen arrived
momentarily, they could attack in unison. The Indians
would scatter, and the 7th would ride them down like so
many rabbits in the grass. This would be his final battle,
and he would be a hero once again.

When it became clear to Custer that the Indians
were not going to flee into the Bighorn Mountains,
that they were not going to simply stand and defend

the village, but that they were going to counterattack en masse with ferocity, courage, and skill, what did Custer feel? What was going through his mind? Was he glad that the Indians would engage in a pitched battle? Was he angry that he might actually have to "work" and fight it out instead of gliding into the village with horse and rifle and announcing that the Indians were captured? The more interesting question, one that will also be debated forever, is whether at any point during the battle, from the first aborted attack at Medicine Tail Coulee to the final desperate fight on Last Stand Hill, was Custer ever, viscerally, afraid? If not for himself, was he afraid for his four relatives that would die with him? Was he afraid for his men? With the quality of command he was leading, and the quality of enemy he was now facing, if Custer was never truly terrified at any point during the battle, then he simply refused to accept what his eyes and ears were telling him.

Custer was leading an outfit that has been called "elite." However, if we closely study the battlefield readiness and combat fitness of Custer's men, what we find is a hodgepodge unit consisting of everything from experienced Civil War veterans to green recruits who had never even once fired their service carbine before riding into battle against the Lakota:

> *"Pictorially, the Seventh was characteristic of...the United States Army—an army sometimes led by cripples and alcoholics, officers who would not be commissioned today or who would be removed from field service. General Oliver Howard, for instance, had just one arm. Gibbon and the myopic Terry both*

were hobbled by Civil War injuries. Grasshopper Jim Brisbin
was rheumatic, frequently resorting to crutches and unable
to mount a horse. Custer's nemesis, General David Stanley,
was notoriously soused on the banks of the Yellowstone and
elsewhere. Reno, Benteen, and countless others seldom were
so impolite as to decline a flask. One gets an impression that
half of the wasichu commandants were physically limited
and/or drunk, to say nothing of their wonderful neuroses and
obsessions—Custer, for example, washing his hands again and
again during the Civil War bloodbath...In addition to senior
officers such as these, various young men were in questionable
shape. Godfrey, at that time a lieutenant, was rather deaf. Lt.
John Crittenden, who transferred from the Twentieth Infantry
and died on the ridge with Calhoun, was blind in one eye.
Lt. Algernon Smith had been so shot up during the Civil War
that he could not raise his left arm above the shoulder or put
on his coat without help. Artists who set out to recreate a
unit of the nineteenth-century United States Army such as
the Seventh Cavalry therefore had a problem. A short-haired
general commanding what might be mistaken for a limping
drunken mob of itinerant farmhands would be altogether
unsatisfactory."

Custer's frequent forays to the East, away from his
command, shed a great deal of light on what he thought
of the Army at this point in his life and what he expected
to face when his command left for Sitting Bull's camp.
Despite Evan S. Connell's damning assessment, some
of Custer's officers and men were combat-hardened
veterans who could have had a greater effect on the unit's
overall performance. Fifteen Medals of Honor were
awarded to soldiers for valor during the battle. Unlike
the Medals of Honor that were shamefully awarded

after the Wounded Knee Massacre, these were rightfully earned. One such recipient, Private Peter Thompson, made multiple trips to the river to gather water for his wounded and dying comrades lying in the makeshift hospital on Reno Hill. Custer's fate was unknown at this point, his command was cut off from Reno and Benteen, and the epic Lakota counterattack had turned the battle into a desperate fight to survive on the hill until the Dakota Column arrived with reinforcements. Many of the soldiers were suffering from severe dehydration due to blood loss and the high ambient temperatures. Anticipating a quick victory, some had carried half-empty canteens into battle. They were pleading and begging, even threatening any able-bodied men to bring them water. Lakota sharpshooters were well aware of the soldiers' predicament and were therefore raking the riverbanks with near-constant gunfire. Thompson promised a wounded friend that he would help him and made the suicidal trip to the river alone, despite being wounded himself. With rounds strafing the water all around him, Thompson worked quickly filling canteens before returning to the safety of the ravine. He repeated this harrowing journey and was shot three times for his efforts. Thompson's bravery, selflessness, and tenacity speak to what might have been had Custer been interested in true leadership, in comprehensively developing and training his command before the battle.

Custer neglected to fill this role, however, instead doing the bare minimum in the way of planning and preparation. He was at the end of his career, and he knew it. His sights were set on politics, the lecture circuit, and various other lucrative opportunities that would secure his future. He likely thought of this venture as a valuable errand, a grand resume builder, to be completed as quickly as possible so he could then refocus on his personal prospects. Frustration with his absenteeism, lack of preparation, and self-involvement finally boiled over when his most trusted ally and friend, General Sheridan, scolded him in a message to President Grant concerning the much-anticipated campaign:

> *"I am sorry Lieutenant Colonel Custer did not manifest as much interest by staying at his post to organize & get ready his regiment & the expedition as he does now to accompany it."*[2]

Seventeen countries were represented under Custer. This was not an army of elites imposing their will and the will of the country on the Lakota. Much of Custer's command consisted of poor immigrants, a former slave, as well as Indian scouts who were all employed to kill the Lakota. Many had no better command of the English language than the enemy they were fighting:

> *"Forty percent of the soldiers in Custer's Seventh Cavalry had been born outside the United States in countries like Ireland, England, Germany, and Italy; of the Americans, almost all of them had grown up east of the Mississippi River. For this decidedly international collection of soldiers, the Plains were as strange and unworldly as the surface of the moon."*[3]

These men came to America not to join the Army, but like most immigrants, they were simply hoping for a better life. Ironically, some even emigrated to avoid conscription in their own nation's military. We have to ask why Giovanni Martini and Charles DeRudio were fighting Crazy Horse and Sitting Bull? Why were Native Italians fighting Native Americans? What could these people possibly have against one another? What the Army offered the most oppressed of its citizenry was employment. The country was in a depression, and they accepted one of the few jobs paying a steady wage. Life in the Army was bleak, but with few options other than striking out for the gold fields, these men had to accept something much different than the American Dream they had hoped for. Soldiers languished in western outposts, suffering from scurvy, malnourishment, boredom, and low pay. The occasional drink or trip to the whorehouse was all many of them had to look forward to. The heroism of the Civil War was over. The Army now pursued an enemy that refused to fight as the South had fought. There were few gallant charges and little in the way of glory. The dynamics of the Indians Wars compelled them to attack villages at dawn, inevitably and sometimes intentionally, killing women and children. There was no honor in it. The Lakota were fighting to defend their families and their way of life. Plagued by a depressed national economy, desertions, disease, harsh discipline, and poor training— the men under Custer were fighting for little more than a paycheck.

This "elite" unit languishing and isolated on the plains, led by a commander with one foot out the door,

preoccupied with his future and life in the East, was
about to attack roughly two thousand warriors who,
man-for-man, comprised the most capable light cavalry
in the country. Unlike the men under Custer, Lakota
warriors were trained for battle as soon as they could
mount a horse:

> *"...From about the age of five, boys were taught the skills
> necessary to be hunters and warriors, such as tracking,
> hand-crafting tools and weapons, close combat, proficiency
> with weapons, and horsemanship. They were mentored one-
> on-one by older males—fathers, uncles, grandfathers, family
> friends—until about the age of sixteen or seventeen. They
> were taken on hunts and military expeditions to observe
> experienced, full-fledged hunters and fighting in the field. By
> their late teens, Lakota males were ready to fulfill their roles
> and take their places as hunters and warriors...Furthermore,
> every fighting man was taught that his commitment as the
> warrior ended only with death, whether it came on the field
> of battle or with old age. Dying in the defense of family and
> home, however, was the highest of honors and the strongest
> legacy to leave behind. It was this kind of fighting man that
> the Seventh Cavalry faced in the valley of the Little Bighorn."[4]*

This meant that the average warrior under Crazy
Horse had years, if not decades, of combat experience
against Indian and white enemies alike. A portion of
Custer's men had rarely fired their carbine, hadn't a clue
how to clear it if it jammed (which it did frequently),
and hadn't received even the most rudimentary training
in Indian warfare. Custer's absenteeism, coupled with
the suspect quality of many of his subordinates and the
prevailing wisdom that Indians couldn't or wouldn't

fight when facing a calvary charge, left much of the command undertrained and overconfident.

Crazy Horse received word of the attack during Reno's charge. Like many Lakota, he was shocked to learn that the Army would advance on such a large village. As arguably the most experienced, successful, and competent warrior in camp, he was expected to lead from the front. He waited, however, to properly prepare himself and ensure his medicine would be strong. Many of his most loyal warriors became impatient and frustrated with the relative calm that descended over their leader as the battle was unfolding. Crazy Horse was often pensive before battle, a characteristic that likely saved his own life and the lives of his men on more than one occasion. One Oglala mentioned that "he didn't like to start a battle unless he had it all planned out in his head and knew he was going to win."[5] Regardless, his men pleaded with him to hurry, but he would not be rushed. When he finally felt that the time was right, his medicine was good, and his pony was ready, he led his men into the fight. He would make up for lost time, however, with one of the most daring and crucial charges of the day, executed at the very moment when Custer's men nearly established a defensive position.

The soldiers had been searching for a viable ford to cross the river and attack the village. They faced fierce resistance from a relatively small number of Lakota warriors from inside the camp. Therefore, they had retreated uphill to a narrow ridge overlooking the valley after several attempts to breach the village had failed. Largely unaware of Reno's fate, this was the first sobering indication for Custer's immediate command that the battle was not going to be an easy one. The Indians were not only standing their ground, but were counterattacking aggressively, and without the timely arrival of Benteen, the situation could deteriorate rapidly.

> *"From a rise [the warriors] heard the gunfire. The soldiers were on a long ridge and moving northward, some of them on foot. Loose horses were everywhere. As they rode closer to the action, it was clear the soldiers were fighting a running battle, and losing. One group, however, seemed to be holding their own, staying together and returning shot for shot at the advancing warriors. Crazy Horse circled farther to the east and then turned south and charged that troublesome group. With warriors behind him, they were able to inflict casualties and scatter the stubborn knot of soldiers."[6]*

Hundreds of warriors were returning from the fight with Reno to join in this new engagement. Custer found himself overwhelmingly outnumbered, with his command spread dangerously thin. He undoubtedly still held out hope that Benteen would arrive, but conditions were such that he could no longer keep his men together fighting as one unit. Instead, Custer was forced to string them out along the ridge, left to defend themselves as he desperately searched for an opening to attack the village from the north in a last-ditch effort to win the battle. Crazy Horse understood that the battle was reaching a tipping point and that severing the commands would likely trigger a rout:

"As pressures mounted to the south, Crazy Horse struck to the north. Extending between Calhoun Hill and the flat-topped knob where Custer and the Left Wing had deployed was a hogback that came to be known as Battle Ridge. For Keogh's Right Wing, this narrow ridge, which extended north like the sharp-edged spine of a gigantic and partially buried beast, was both a bulwark against the Indians and a potential pathway to Custer and the Left Wing. By riding his pony through a slight gap in the forty-yard-wide ridge, Crazy Horse managed singlehandedly to break the Right Wing in half."[7]

Crazy Horse's charge across open ground, against an entire company, was more than a suicidal dash. He neutralized the soldiers' only chance for an organized defense before they were all killed. He single-handedly cut off forty men from the rest of the command during the most critical point in the battle. Just as it seemed Custer's men might establish a defensive position and turn the tide, Crazy Horse broke through the line and severed the commands. "Crazy Horse was the bravest man I ever saw...," marveled the Arapaho Waterman. "All the soldiers were shooting at him, but he was never hit."[8] The soldiers then fell into a full retreat, having given up on any attempt to hold ground as they waited for reinforcements. Crazy Horse and his men continued on to Last Stand Hill and helped crush what was left of Custer's command.

Crazy Horse's charge at the Battle of the Little Bighorn was perhaps the most singular act of defiance in the history of Native-Euro relations. With that assault, Crazy Horse reified his humanity in the midst of battle.

He proved that one competent individual, with agency, can check the immense power of the state. Crazy Horse performed consistently well in combat, and he led from the front in three great battles against the US Army: The Fetterman Battle, The Battle of the Rosebud, and The Battle of the Little Bighorn. But he was much more than a talented war leader. He embodied the four Lakota virtues of generosity, courage, fortitude, and wisdom and carried them out in his daily life; he often ensured that widows, the elderly, and other persons in need were cared for and well-provisioned:

> *"Crazy Horse wasn't perfect but he was generous with his material goods and his efforts on behalf of others. He demonstrated courage time and again on and off the battlefield. His fortitude enabled him to hang on to his values, beliefs, and principles during a time of traumatic change for the Lakota, and he worked to acquire wisdom, realizing that it comes from failure as well as success. He was much the same as other Lakota men of his day, indeed the same as most Lakota men of the nineteenth century. Like them, Crazy Horse was many things and fulfilled many roles. He was a son, husband, brother, father, and teacher. He was a crafter of weapons and tools, a hunter and tracker, horseman, scout, and fighting man, to list a few. He was also a deep thinker, a shy loner, a fierce defender of all that he held dear, a keen observer, a rejected suitor, a moral person, a family man, and a patriot..."*[9]

Crazy Horse clung to his world even as it was crumbling around him. He understood what reservation life had in store for him and knew he was ill-suited for the politics and in-fighting that would take place amongst the leaders put in place by the US government. Like Custer, he was a human being who made choices on and off the battlefield that greatly impacted his people and his way of life. Unlike Custer, he was more consistent, more disciplined, and more selfless. He had neither the political aspirations nor the power-seeking ambitions of Custer and other Lakota leaders at the time. He was a man who retained his agency and independence until starvation forced him to capitulate. And his epic charge at The Battle of the Little Bighorn remains one of the most salient acts in our nation's history of what a properly motivated individual can achieve when faced with the loss of their freedom. If Custer was the harbinger of change and the catalyst for monomaniacal progress, Crazy Horse was the vanguard against the onslaught of modernity, the standard-bearer for tradition and self-determination. His bravery in battle, fidelity to his people and way of life, and commitment to his own humanity and freedom, are some of the most instructive examples of this period that we can learn from as we navigate our own traumatic change in 21st-century America.

EPILOGUE

It is easy for me to imagine that the next great division of the world will be between people who wish to live as creatures and people who wish to live as machines. **—Wendell Berry**

The historical precedent of the US Government's subjugation of the Lakota and other tribes through the removal of the buffalo and consequent protein dependence provides a template for which we will suffer the same fate. Washington realized that fighting the Lakota was the least efficient and most expensive method of oppression. Generals Sheridan, Sherman, and others confessed from time to time that defeating the Lakota militarily would probably be possible in the long run but would require an untenable amount of blood and treasure. Sherman even once lamented that fifty well-trained warriors could checkmate three thousand US soldiers. Not to mention that the nature of Indian fighting often resulted in an excessive amount of non-combatant deaths that the voting public would be unlikely to tolerate long-term.

The demise of the buffalo was the silver bullet the government had been searching for since the end of the Civil War. Battlefield success matters little if an army cannot be fed. Crazy Horse and Sitting Bull could resist alcohol, coffee, sugar, and clothing, if they so desired. But they, like us, needed protein to survive. When the buffalo fell to near extinction levels, the Lakota effectively became protein hostages of the United States Government.

What does this mean for us? After all, we are not fighting on the battlefield or living off the land, not looking for buffalo to kill or horses to steal in order to perpetuate our culture and way of life. If we look closely at the government's attempted extermination of Lakota culture, we find that Crazy Horse and Sitting Bull did not surrender because they were defeated militarily. They did not surrender because they felt that the American way of life was better. And they certainly did not surrender because they wanted to. In the final analysis, two of the most powerful men in the country gave up everything they had because the powers-at-be had eliminated their ability to provide for themselves. They were starving, and they were simply out of options. The generals and the government, as a whole, understood that the easiest (and cheapest) way to control a people and a culture is to control their food source. The Lakota could not eat cash, gold, guns, or bullets, and neither can we. Once our protein source has been centralized, controlled, regulated, and rationed, we will also become protein hostages. Therefore, we must retain as many alternatives regarding our food procurement as possible.

We are being increasingly convinced that the best option for obtaining food is to simply push a button on an app or a website, and then wait for it to magically appear on our doorstep. It is convenient, cheap, efficient, and climate-friendly. We need not leave our homes. And many of us are buying into this—we are getting exactly what we want. We are not thinking about the options that we are losing and what this means for our agency. If readers remember nothing else about this book, I hope they remember one big takeaway: As we transition into a fully-automated, digital, and curated world and life experience, we, as individuals, a society, and a country, must retain as many options as possible regarding our food procurement in order to preserve our freedom, agency, and humanity. If we retain autonomy over our food, then we, not Tech, will dictate and control our future. We must empower each consumer to decide what is best for themselves and their families. Their available options must consist of more than just pushing a button on an app or a website and waiting for UPS to deliver what they have so quickly and conveniently ordered. If this is all we have, then we will be compelled to conform to the entity controlling the flow of food, compelled to conform to the political, societal, and cultural norms that said entity espouses. What might happen if we decide one morning that we are running low on food and fire up our smart devices to order more, only to find that we are unable to do so? Is this outage due to a Russian cyber-attack, a processing facility fire, a pandemic, or just another so-called "black swan" event which will continue to occur with alarming frequency but nevertheless be called a black swan event by Tech and Big Ag? Why should we accept a system that can be completely disrupted by a nefarious hacker a continent away who strikes a few keys one morning while he sips his coffee and checks his crypto account? Those of us who have accepted this system might be induced to panic. Others who are participating in an alternative food system, who have a freezer full of real food and real protein, will likely yawn and watch the spectacle with tired eyes.

As Americans, we tend to have perpetually short memories. This disruption could never happen in the US, we might say. This is the land of plenty, and there will always be food, and no actor, public or private, foreign or domestic, could ever deprive us. But we have already experienced a dry-run, so to speak. The Covid-19 pandemic laid bare what can and will happen to our food supply if we accept a techno-centric, consolidated food system as our only option. Who thought we would ever see bare shelves in this country, the wealthiest on the planet? No beef, no chicken, no pork, no milk, no eggs. Forget the toilet paper. You won't have any need for it if you have nothing to eat.

The great irony, of course, is that there was plenty of food available—plenty of beef cattle, hogs, sheep, dairy cows, and chickens laying eggs. But because of the logistical nightmares the pandemic triggered, these animals could not be slaughtered, processed, packaged, and presented to the consumer. Many farmers were forced to euthanize hogs, dump milk, and cull cows. What resulted was one of the most horrifying losses of food this country has ever seen. Most of us wouldn't have had the first clue of what to do if a farmer dropped off a steer or hog in our backyard. However, I believe it would have been quite beneficial for us as citizens if we had been forced to try. I think we would have been very surprised at what we are capable of given the proper motivation. People would have thought that it was impossible, beyond their abilities, better left to experts in inspected, automated processing facilities comfortably removed from their sphere. The average person could never butcher an animal, you might say. The Lakota butchered buffalo on a regular basis. Are they so different from you?

Imagine a family of four—mother, father, and children attempting to butcher a hog in their backyard for the first time. It would be inefficient and bloody. The bladder and intestines would likely be punctured, and much of the meat would end up wasted. But the yield would still be well over a hundred pounds of pork to be shared with relatives, neighbors, and friends. After they finished, the family would be amazed at what they accomplished. And they would have much less trepidation and anxiety when the next food shortage rolls around. Perhaps they would be less inclined to listen when they read an online post telling them that killing and eating animals is wrong and antiquated and disgusting and that pushing a button and waiting for synthetic protein to arrive on their doorstep in a neat package is in their best interests if they want to be good people and good citizens.

The pandemic was not punishment for wayward behavior or non-compliance; it was a tragedy of epic proportions. However, it provides an instructive example of how quickly and catastrophically our food supply can be disrupted, or even completely arrested. Is it such a great logical leap to envision a scenario in the future, a future when every step along the supply chain is automated and controlled by a singular central entity (whether we realize this or not), that the flow of our food could be temporarily or permanently halted? The pandemic demonstrated the systemic fragilities and weakness inherent within 21st-century American food production. We are being told that the solution to this existential crisis is not less centralization and consolidation, but *more*. We should support, financially and politically, companies that produce synthetic protein because these are the healthiest, cheapest, and most environmentally-friendly products being offered. Everything else is inefficient, costly, and contributes to climate change. Therefore, all citizens should patronize only those companies that manufacture synthetic meat because it is the "right" thing to do. Anyone who does otherwise is facilitating the murder of innocent beings and the death of our planet.

Industrial agriculture certainly has its faults, and a

discussion on that industry is beyond the scope of this book. But I am not advocating for industrial agriculture. I am not advocating for more concentrated animal feeding operations (CAFOs). I am not advocating for more centralization of our food supply under any circumstances, be it real or synthetic. I am advocating for *de-centralization*, for a citizen's right to participate in an alternative food system, to procure their own food through hunting, fishing, gardening, and animal agriculture, or to patronize independent, regenerative family farmers if they are unable to do so themselves.

Crazy Horse and Sitting Bull consistently refused to listen to the lies espoused by the US government and the US Army. Many others did listen, however. Promises of annuities, rations, hunting rights, and land cessations often went unfulfilled or, in many cases, completely reneged on. Treaties were made and broken only for more to be made and broken again as the needs of the US government and the American people changed. For many native peoples who had believed in the treaties, it was too late. Their freedom had already evaporated, never to return. Now, they had to abide by the societal and cultural norms the government espoused. It should come as no surprise that these norms consisted of a central tenet: surrendering all agency to that whom provides all wants and satisfies all needs.

We will be forced to surrender as well. Our own needs will change. Screens will replace our world. If it can't be accomplished through a screen, it is not good for us, not safe, healthy, or right, and we should avoid it.

You could get killed riding your bike on the road, good citizen—much safer to stare at the screen on the wall while riding a Peloton where we can ~~monitor~~ help you. Eventually, we will become life-long inmates in a prison we constructed. We will be "institutionalized" through and through, unwilling and unable to walk out into a world of risk, uncertainty, and freedom. We will gladly eat what we are told to eat, buy what we are told to buy, read what we are told to read, and believe what we are told to believe.

Maybe some of us would like to prevent this outcome. How do we do it? By actively limiting the march of technology on our lives, we can mitigate what is exacted from us, and we can retain more of our freedom, agency, and sanity. But what does that mean in our day-to-day lives? What steps can we take right now? The simplest approach is to connect with and participate in the natural world as often as possible which will reinforce and perpetuate a connection with your own humanity as well as, and just as critically, provide you with independent sustenance. The Lakota and other tribes have long believed that in order to maintain spiritual and mental health, an individual must remain connected to the land and to nature. If one becomes removed or disconnected, spiritual and mental illness will result. Although this philosophy (oversimplified for the purposes of this discussion) is thousands of years old, it is especially relevant to 21st-century existence. Our societal woes have many causes, but our increasing screen time and self-destructive choices are leaving us chronically ill emotionally and spiritually. 65% percent of American adults play video games.[1] Video games. People in the prime of their lives, with families and careers, choose to spend their leisure time in a virtual world while less than 5% engage in hunting and consequent connection with the natural system.

Hunting is a religion when it is done well and done right. You feel a connection to the animal and to the cycle of life and death. You are an active participant in this cycle, as humans have been for two million years. It truly is a spiritual experience. But just like our own lives, hunting is unpredictable and difficult. It's bloody, violent, and can be brutal at times. Animals often twitch, gasp for air, choke, and spasm. There is always the chance of a bad hit, a wounded animal eluding us to suffer and taking longer to die than necessary. These events are rare, but they do happen because hunting is a real experience with no guarantees or Tech-controlled curation. The animal's death should happen quickly when done well, but we still witness it. A heart was beating, lungs were inflated, and blood was flowing through a sentient being just seconds before. Taking that life is a sacred act, and our participation in the cycle of life and death is also sacred. Despite the remorse we may feel watching a life end that we caused, this system is real, natural, and honest. The system that will replace it surely will not be.

It should not be assumed that I am advocating every able-bodied US citizen participate in hunting. I understand that not everyone desires to hike into a wilderness area, kill an animal, butcher it, and pack out hundreds of pounds of protein on their backs. It's extremely physically demanding, time-consuming, and many may simply feel that they do not have the mental, emotional, or spiritual disposition to take an animal's life. There is no harm in that. As someone who is passionate about nature and hunting, I can tell you that I fully support hunting seasons. I would never want to, really would be unable, to hunt year-round. An individual must be in a certain place emotionally and spiritually to take the life of a wild animal. This mental "space" cannot be maintained 24/7, 365. A person must gradually work their way into it and out of it. There is an ebb and flow in our psyche between life-taker and life-protector, a duality that exists in all of us. From September through December, I want nothing more than to provide wild sustenance for myself through fair-chase traditional bowhunting. By January, I have lost all desire to end an animal's life. I transition back to a conservationist, wanting to watch deer and other wildlife and protect and preserve the natural places in which they live.

What we need, and what must be maintained at all costs, is a percentage of our citizenry actively engaged in hunting, fishing, gardening, and animal agriculture—the more, the better. In the world I envision, soccer moms will butcher deer in their suburban backyards while their peers languish on Facebook. *Call of Duty* warriors will trade in their headsets for a bow or rifle and begin to fight for a real life. Children will milk cows, raise chickens, and plant gardens. They will learn to rely on themselves, not the device in their hand. If this is the case, then the entirety of the American people benefit. Even those who would like nothing more than to recline in front of a screen every free minute they have and order food from Amazon. Everyone will benefit because this "rogue" faction that hunters and farmers represent will create a bulwark against a comprehensive usurpation of human freedom. As long as some of us are exercising our right to participate in an alternative food system, to opt out, so to speak, Tech will be held in check.

Crazy Horse and other traditionalists represented a "rogue" faction as well. Those Indians already living on reservations benefitted from his autonomy. His independence forced the US government to treat the agency Indians better than they might have with some degree of fairness and honesty. These terms, of course, are relative. The government's treatment of native peoples in the 19th century was rarely fair or honest, but individuals who refused to capitulate forced the government to keep some promises in an effort to convince the agency Indians that they were better off on reservations as opposed to joining Crazy Horse. Custer was defeated largely because Sitting Bull's village was reinforced with warriors already living on reservations, already under the government's control, who absconded to join the village in the days ahead of the battle. Custer and the other officers afield knew little of this exodus. General Sherman's first verbalized thoughts after learning of the disaster were not related to Custer's supposed recklessness or incompetence. Instead, he expressed utter disbelief that there were enough Indians out there to wipe out an entire command. Where had they come from? These Indians were only able to leave the reservation because an alternative still existed. Sitting Bull and Crazy Horse represented that alternative. We must do the same. We must continue to reinforce our agency and humanity to maintain another option, an alternative to the digital reservations where many of us so willingly reside.

The traditionalists were constantly courted by the US government to lure them onto reservations. Social Media Influencers, the Lead Cows, are doing the same thing for Tech. They are courting us in an attempt to convince us to spend more and more time on digital reservations until we are unwilling or unable to leave. Once the traditionalists had surrendered, there was little incentive for the government to behave well. The Wounded Knee Massacre was the inevitable conclusion, an epic and horrifying testament of what happens when a people and a culture have lost all agency and freedom and simply run out of options against the state.

Many people disparage hunting and animal agriculture. It is anachronistic, barbaric, and unhealthy for all involved and should be eliminated through regulation and consequent criminalization. Once these activities are gone, what will take their place? What system will we become a part of? Are we better off placing all of our needs in one basket and purchasing lab-grown meat and produce from companies who see us as nothing more than profit centers—vessels that must be filled with the cheapest, most efficient, and cost-effective "food" on the planet? Margins must be made, and investors must be paid. No company survives on "thank you." We certainly cannot fault anyone for starting a business and trying to turn a profit. This is America. But is this what we want when it comes to our food—what will feed to our children, our spouses, and ourselves? Do we want synthetic sustenance to go along with a synthetic life, a synthetic world, a synthetic future?

As a culture, we have become accustomed to outsourcing everything. We outsource child-care, education, grocery shopping, cooking, cleaning, home and lawn maintenance, dog walking, transportation, and more. Increasingly we have begun outsourcing our thinking. Whenever we are presented with a dilemma, a new concept, or an unfamiliar idea, we don't attempt to educate ourselves. It takes too much time, it's too much work. Sloth is no longer a deadly sin; it's a virtue. It's so much easier just to listen to our own village. If I'm a Democrat, I listen to other Democrats. If I like Rush Limbaugh, he will surely tell me what to do and what to think. If you are not one of us and you do not think the way we do, then there is something fundamentally wrong with you. Your facts are wrong, of course, but more importantly, you are somehow compromised. You have neither the mental faculty nor the intestinal fortitude to think and do what we know to be "right." Much of this information is transmitted through social media—Facebook, TikTok, and Twitter—who have no other incentive than to hold our attention as long as possible. We have become empty vessels, thoroughly appropriated, and simply waiting for our instructions.

I once heard an NPR guest state that "to be human is to be connected." Simple, yet quite profound, and the question we will have to answer, the question we are answering every waking minute with every decision we make today, tomorrow, next week, next month, and next year is connected to what? What future do we want? As we transition from one way of life to another, how do we ensure that our lives and the lives of our children and grandchildren will be better, more fulfilling? How can we ensure they will contribute more to society in the next century than Facebook, Amazon, or Twitter? How can we prevent subsequent generations from living enslaved lives of comfort and ease behind a screen all day, every day? After all, who among us envies the primates behind the glass?

Sitting Bull told his people to adopt what was good in white culture and discard what was bad. Good advice, to be sure, but only possible if one still had that option, if one still retained agency. Once the Lakota were forced onto reservations through protein dependence, how could they heed such advice? It probably never occurred to Sherman, Sheridan, or Grant that instead of fighting the Indians and slaughtering buffalo en masse, wouldn't it have been so much easier, faster, and cheaper to simply convince them that killing buffalo was morally wrong? That they should feel ashamed of themselves for murdering sentient beings, that it was wreaking havoc on their own psyche, and that their children would grow up to be deranged sociopaths. That probably would have garnered a good laugh from everyone involved, and the generals surely would have known that such an argument and such philosophy was ridiculous, futile, and hypocritical. The buffalo was the center of the Lakota world. It provided spiritual and physical sustenance, as well as tools and materials for daily living. The Lakota needed the buffalo to live. Both were locked in the natural cycle of hunter and hunted, life and death. To ask or even hint that such a system should be abandoned because it was "wrong," "violent," or "unnecessary" would have been absurd. Besides, who determines what is wrong, or violent, or unnecessary? Who ultimately benefits from a people believing the cultural norms espoused by the powers-at-be? What code of morality is being employed, and how malleable is it? The Lakota surely would have asked similar questions before they dismissed such an idea out-of-hand. Why should we react any differently?

We must also heed Sitting Bull's message. We must maintain a selective-adoption-of-technology mindset, incorporating what is good and discarding what is bad. We cannot reflexively accept every Tech-curated "solution" as automatically beneficial to society without even a shred of skepticism. We must think for ourselves and determine what is best for each person's right to agency, freedom, and self-determination. If I have learned anything in my time on this planet, it's that I do not want other individuals or entities to curate my cognition. Elon Musk, Mark Zuckerberg, Jeff Bezos, Jack Dorsey, Bill Gates, and all the rest are developing the most advanced technology in the history of the planet, we are told. They will give us automation, AI, virtual worlds, and work-free lives full of perpetual ease and happiness. We are so lucky to be living at this time in history, we should be grateful, and we should never question whether these innovations will truly benefit us. We should never stop to think that every man, woman, child, and transgender person already possesses the most advanced technology humans have ever developed. Regardless of where they live, their skin color, their socio-economic status, their gender, or their sexual orientation, they already own the original "smart device" that Musk and Zuckerberg, and Gates could only dream of inventing. And they got it for free. It's what sitting between their ears. We need to use this "technology." We need to think critically about what we are being told, what we are being sold, and who is going to ultimately benefit. These titans of technology have not yet been able to out-develop the human brain, but they are trying like hell. And one day, they just might succeed in making it obsolete. Thinking will be considered "work," something to be avoided at all costs because we have advanced past this unenlightened, lowly stage in our evolution. Our cognition will be thoroughly curated, and then our humanity will finally be usurped once and for all. We will become, as Wendell Berry states, simply machines, empty vessels waiting to be filled.

An astute person will now interject and say that this book is one big attempt to convince its readers to act or think or behave in a certain way. And I would agree, wholeheartedly. I am attempting to convince every reader to preserve hunting, angling, gardening, and animal agriculture. I would also encourage those same individuals to educate themselves regarding the anti-hunting, anti-farming, and animal-rights agenda and philosophies. Use your brain, inform, and educate yourself on which path will most benefit you as an individual, your children, your grandchildren, and most importantly, wildlife and wild places. Then vote accordingly. I would rather meet a well-informed, intellectually curious, open-minded opponent, than a zombie who supports hunting simply because he saw a pretty girl holding a rifle on Instagram.

Of course, like anything that is real in life, you will eventually have to stop reading and thinking and listening and watching, and you will have to participate, actually have to put forth the time and effort and work to truly know what a thing is about. And so it is with hunting. My hope is that one or two readers will do what I am going to suggest in the following paragraphs. The rest will write me off as a redneck, weirdo, pariah, and criminal. They will say it is too dangerous, violent, and really a waste of time. Then they will return to Facebook to read about the lives other people are living.

For the individuals still reading, here is what I propose: Contact your state conservation department, attend the necessary classes, and complete the coursework to obtain your deer-hunting license. Purchase a rifle, take a firearm safety course, and practice as often as possible. Listen to other hunters and marksmen. Learn from them and heed their advice. Find a mentor and accompany this person afield. You will learn a lot if you are willing and patient and committed. You will eventually be a part of a successful hunt. You will feel lots of emotions running through your body upon this experience. It will be a reawakening, a retroactive baptism into your own humanity. You will understand that procuring your own protein is possible, within your abilities, if only you put forth the time and effort.

However, to achieve what will be the most lasting, addictive, salient, and beneficial experience concerning your own independence and agency, you will eventually have to proceed alone. Find an area to hunt before the season. Look for trails, rubs, scrapes, scat—any sign that deer are around. Think about what you have learned and the advice you have been given. Construct a blind on the ground. Hunt as often as possible. Tell your spouse, your friends, and your family what you are doing. Take careful notice of their reactions. Write them down. You will eventually see a legal deer, but this process could take years. You will think you are simply wasting your time.

When you finally see a deer within range, move slowly, and put the crosshairs over the lungs. Take a breath and hold it. Squeeze the trigger, and watch the deer. Wait. Wait some more. Then pick up the blood trail and follow it. As you approach the deer, think about what you have done and why you have done it. Field-dress the deer and drag it out of the woods. Hang it on a gambrel, skin, and butcher it. You will be shocked at how difficult it is, how long it takes, and the amount of work involved. You will be covered in blood, hair, and the musky stench of deer. You will curse me for having ever gotten you into this. Should you be cursing me or cursing yourself for this reaction? Does this mean anything for your own journey as a human, for how far you have removed yourself from real life and real experience?

As soon as possible, take the backstraps and cook them, preferably over an open fire or a bar-b-que pit. Eat them alone. Think about the long process that has gotten you to this point. Think about why you have killed this animal and what it means for your life. At that moment, you will experience a feeling unlike any you have ever known. It will wash over you and increase in intensity. A natural high—a dopamine ambush triggered by participation in the most human of activities, a desire that has been there all along, repressed and forgotten, buried deep within your psyche and your soul. That feeling is freedom. From start to finish, you have procured your own food. The tremendous amount of time and effort and work that you have expended has set you free. You hunted, killed, butchered, and consumed an animal for the first time in your life. Now read your friends' and family's reactions that you had written down. You will never want to buy meat from the store again. You will want to, as often and practical as possible, provide food for yourself, your spouse, and your children that which you have hunted, killed, processed, and cooked. No one will ever care as much about what goes into their bodies as you do.

These game animals are not products with profit margins baked in. They have not been mass-produced with relentless efficiency and cost-cutting; they are not commodities to be bought and sold to make a return for an investor. They exist within the sacred cycle of life and death, of beings consuming other beings, a system that we have always been a part of and always will be. This homecoming of sorts will be the most transformative experience of your life. All other aspects of modern living—social media, apps, the metaverse—will pale in comparison. You will have experienced the real thing and you will crave this feeling, this high above all others—a spiritual itch you can't quite scratch until the next hunting season arrives. You will finally understand why some people go to church, and others go hunting.

It seems that as humans, we have become obsessed with progress, regardless of the costs. We are relentlessly focused on making our lives easier, safer, more predictable, and "better." And the underlying assumption is that this is a moving target, a goal we are always chasing, one that none of us will accomplish before we die, our work to be picked up by the next generation and carried forward. Is there a point at which our maniacal pursuit of progress reaches critical mass? Will the costs eventually be greater than the benefits? Is there a line at which progress for the human race becomes self-defeating? Is our enslavement to automation a foregone conclusion? Are we creating more problems than we are solving? Are we evolving ourselves out of existence? Will the human brain eventually become obsolete? Are we working diligently and tirelessly to dig the grave of humanity all the while congratulating ourselves for developing an ever more efficient, modern, and artistic spade?

Amazon, drones, automation, and AI will be the gods that we will pray to in the future, already are praying to. In time, it will be as if our way of life never existed at all, an archaic step in human evolution, that future generations will look upon with fascination and intrigue, but with no hint of envy, just as we look upon the traditional Lakota way of life today. We must be cognizant of the price of progress, must be more cautious and diligent in the future. We must continue to maintain a selective-adoption-of-technology mindset. We must strike a balance with Tech in our lives to retain our humanity. We must remain vigilant against these incursions on our agency, must fight to retain our freedom and sanity every waking minute of every day until we die and leave the battle to our children and grandchildren.

Perhaps there will eventually be a generation that completely capitulates to Tech and an automated world. They will eat what they are told to eat, buy what they are told to buy, read what they are told to read, and believe what they are told to believe. Their cognition will be completely curated, they will have abdicated all freedom and agency, and they will be the happiest people that ever lived because they will be high all the time. The digital reservations will be thriving, with more created every day. There will be a small, independent faction left, of course, as there always is—the unfortunate souls who refused to accept what was best for them. These people will be called the traditionalists, or the hold-outs, the Luddites, the rednecks, the weirdos, the pariahs, the criminals. They will eventually gather what's left of their village and head into the woods one last time. They will cut their bows from the wood and harden them over fire. They will sharpen their arrowheads until they cut through bone. In the final moments, in the deepest recesses of their hearts, they will feel a stirring. They will ask themselves—as the last bastion of people who are still fully committed to their own agency, who still desire a life of work, uncertainty, and freedom, largely independent of Tech, without curation and full of promise, after 2 million years—is this the final chapter? Is this humanity's Last Stand? Has the human race finally evolved itself out of existence? Have we finally realized the true price of our epic obsession with progress? And if so, for the few human beings remaining, could there really be a better time to die?

NOTES

Chapter 2: Citizen Helpless

1. Marshall, David Weston. *Mountain Man: John Colter, the Lewis & Clark Expedition, and the Call of the American West*. New York: The Countryman Press, 2017. 132-133.

2. Diamond, Jared. *The World Until Yesterday: What Can We Learn from Traditional Societies?* New York: Viking Penguin, 2012. 168-169.

Chapter 7: Cast the First Stone

1. "Mine Safety and Health Administration: Coal Fatalities for 1900-2021." United States Department of Labor. https://arlweb.msha.gov/stats/centurystats/coalstats.asp.

2. "New Studies Confirm a Surge in Coal Miners' Disease." National Public Radio. https://www.npr.

org/2018/05/22/613400710/new-studies-confirm-a-surge-in-coal-miners-disease. 05/22/2018.

3. "Black Lung Rate Hits 25-Year High In Appalachian Coal Mining States." National Public Radio. https://www.npr.org/2018/07/19/630470150/black-lung-rate-hits-25-year-high-in-appalachian-coal-mining-states. 07/19/2018.

4. Schiffman, Richard. "A Troubling Look at the Human Toll of Mountaintop Removal Mining." Yale Environment 360. https://e360.yale.edu/features/a-troubling-look-at-the-human-toll-of-mountaintop-removal-mining.

5. "What is US electricity generation by energy source?" US Energy Information Administration. https://www.eia.gov/tools/faqs/faq.php?id=427&t=3.

Chapter 8: A Helping Hand

1. Semuels, Alana. "'They're Trying to Wipe Us off the Map.' Small American Farmers are Nearing Extinction." TIME Magazine. https://time.com/5736789/small-american-farmers-debt-crisis-extinction/11-27-2019.

2. Chadde, Sherman and Wedell. "Midwest farmers face a crisis. Hundreds are dying by suicide." USA Today. https://www.usatoday.com/indepth/news/investigations/2020/03/09/climate-tariffs-debt-and-

isolation-drive-some-farmers-suicide/4955865002/.
03-09-2020.

3.   Gates, Bill. *How to Avoid a Climate Disaster: The Solutions We Have and the Breakthroughs We Need.* New York: Alfred A. Knopf, 2021. 129.

4.   Temple, James. "Bill Gates: Rich nations should shift entirely to synthetic beef." MIT Technology Review. https://www.technologyreview. com/2021/02/14/1018296/bill-gates-climate-change-beef-trees-microsoft/. 02-14-2021.

Chapter 9: Bees in the Hive

1.   Stiles, T.J. *Custer's Trials: A Life on the Frontier of a New America.* New York: Alfred A. Knopf, 2016. 281-282.

2.   Stiles, T.J. *Custer's Trials: A Life on the Frontier of a New America.* New York: Alfred A. Knopf, 2016. 77.

3.   Battle of Antietam. History of Battle of Antietam. American Battlefield Trust. www.battlefields.org/learn/battles/civil-war/antietam.

4.   Connell, Evan S. *Son of the Morning Star: Custer and The Little Bighorn.* New York: North Point Press, 1984. 267.

5.   Connell, Evan S. *Son of the Morning Star: Custer and The Little Bighorn.* New York: North Point Press, 1984. 142-143.

6. Connell, Evan S. *Son of the Morning Star: Custer and The Little Bighorn*. New York: North Point Press, 1984. 142-143

7. Stiles, T.J. *Custer's Trials: A Life on the Frontier of a New America*. New York: Alfred A. Knopf, 2016. 100.

8. Stiles, T.J. *Custer's Trials: A Life on the Frontier of a New America*. New York: Alfred A. Knopf, 2016. 104-105.

9. Merington, Marguerite. *The Custer Story: The Life and Letters of General George A. Custer and His Wife Elizabeth*. New York: Barnes & Noble, 1994. 159.

10. Utley, Robert M. *Custer: Cavalier in Buckskin*. Norman: University of Oklahoma Press, 2001. 40.

Chapter 10: Bravest Man I Ever Saw

1. Connell, Evan S. *Son of the Morning Star: Custer and the Little Bighorn*. New York: North Point Press, 1984. 361-362.

2. Stiles, T.J. *Custer's Trials: A Life on the Frontier of a New America*. New York: Alfred A. Knopf, 2016. 438.

3. Philbrick, Nathaniel. *The Last Stand: Custer, Sitting Bull, and the Battle of the Little Bighorn*. New York: Viking Penguin, 2010. XX.

4. Marshall III, Joseph M. *The Day the World Ended at Little Bighorn*. New York: Viking Penguin, 2007. 80-81.

5. Connell, Evan S. *Son of the Morning Star: Custer and the Little Bighorn.* New York: North Point Press, 1984. 63.

6. Marshall III, Joseph M. *The Day the World Ended at Little Bighorn.* New York: Viking Penguin, 2007. 73-74

7. Philbrick, Nathaniel. *The Last Stand: Custer, Sitting Bull, and the Battle of the Little Bighorn.* New York: Viking Penguin, 2010. 268.

8. Philbrick, Nathaniel. *The Last Stand: Custer, Sitting Bull, and the Battle of the Little Bighorn.* New York: Viking Penguin, 2010. 268.

9. Marshall III, Joseph M. *The Journey of Crazy Horse: A Lakota History.* New York: Viking Penguin, 2004. XXI.

Epilogue

1. 2022 Essential Facts About the Video Game Industry. Entertainment Software Association. 2022 Essential Facts About the Video Game Industry - Entertainment Software Association (theesa.com)

BIBLIOGRAPHY

Anglin, Ronald M. and Larry E. Morris. *The Mystery of John Colter: The Man Who Discovered Yellowstone.* Lanham: Rowman & Littlefield, 2014.

Berry, Wendell. *The Unsettling of America: Culture & Agriculture.* San Franciso: Sierra Club Books, 1996.

Connell, Evan S. *Son of the Morning Star: Custer and the Little Bighorn.* New York: North Point Press, 1984.

Cozzens, Peter. *The Earth Is Weeping: The Epic Story of the Indian Wars for the American West.* New York: Alfred A. Knopf, 2016.

Custer, George Armstrong. *My Life on the Plains.* Arcadia Press, 2017.

Deloria Jr., Vine. *Custer Died for Your Sins: An Indian Manifesto.* Norman: University of Oklahoma Press, 1988.

Diamond, Jared. *The World Until Yesterday: What Can We Learn from Traditional Societies?* New York: Viking Penguin, 2012.

Donovan, James. *A Terrible Glory: Custer and the Little Bighorn—The Last Great Battle of the American West.* New York: Little, Brown and Company, 2008.

Drury, Bob and Tom Clavin. *The Heart of Everything That Is: The Untold Story of Red Cloud, an American Legend.* New York: Simon & Schuster, 2013.

Erdoes, Richard and Alfonso Ortiz. *American Indian Myths and Legends.* New York: Pantheon Books, 1984.

Gates, Bill. *How to Avoid a Climate Disaster: The Solutions We Have and the Breakthroughs We Need.* New York: Alfred A. Knopf, 2021.

Gates, Bill. *How to Prevent the Next Pandemic.* New York: Alfred A. Knopf, 2022.

Gwynne, S.C. *Empire of the Summer Moon: Quannah Parker and the Rise and Fall of the Comanches, the Most Powerful Indian Tribe in American History.* New York: Scribner, 2010.

Hamalainen, Pekka. *Lakota America: A New History of Indigenous Power.* New Haven and London: Yale University Press, 2019.

Hedren, Paul L. *After Custer: Loss and Transformation in*

Sioux Country. Norman: University of Oklahoma Press, 2011.

Marshall, David Weston. *Mountain Man: John Colter, the Lewis & Clark Expedition, and the Call of the American West*. New York: The Countryman Press, 2017.

Marshall III, Joseph M. *The Day the World Ended at Little Bighorn*. New York: Viking Penguin, 2007.

Marshall III, Joseph M. *The Journey of Crazy Horse: A Lakota History*. New York: Viking Penguin, 2004.

Marshall III, Joseph M. *The Lakota Way: Stories and Lessons for Living*. New York: Viking Compass, 2001.

McMurtry, Larry. *Crazy Horse: A Life*. New York: Penguin Group, 1999.

Merington, Marguerite. *The Custer Story: The Life and Letters of General George A. Custer and His Wife Elizabeth*. New York: Barnes & Noble, 1994.

Miller, David Humphreys. *Custer's Fall: The Indian Side of the Story*. Lincoln: University of Nebraska Press, 1985.

Petersen, David. *Heartsblood: Hunting, Spirituality, and Wildness in America*. Durango: Raven's Eye Press, 2010.

Philbrick, Nathaniel. *The Last Stand: Custer, Sitting Bull, and the Battle of the Little Bighorn*. New York: Viking

Penguin, 2010.

Rodgers, Diana, and Robb Wolf. *Sacred Cow: The Case for (Better) Meat.* Dallas: BenBella Books, Inc. 2020.

Sandoz, Mari. *Crazy Horse: The Strange Man of the Oglalas.* Lincoln: University of Nebraska Press, 1992.

Slotkin, Richard. *The Fatal Environment: The Myth of the Frontier in the Age of Industrialization 1800-1890.* Norman: University of Oklahoma Press, 1998.

Stiles, T.J. *Custer's Trials: A Life on the Frontier of a New America.* New York: Alfred A. Knopf, 2016.

Utley, Robert M. *Custer: Cavalier in Buckskin.* Norman: University of Oklahoma Press, 2001.

Printed in the USA
CPSIA information can be obtained
at www.ICGtesting.com
CBHW052251081123
1742CB00003B/5

9 798218 277840